Renewable Energy Crash Course

Eklas Hossain · Slobodan Petrovic

Renewable Energy Crash Course

A Concise Introduction

 Springer

Eklas Hossain
Oregon Institute of Technology
Klamath Falls, OR, USA

Slobodan Petrovic
Oregon Institute of Technology
Wilsonville, OR, USA

ISBN 978-3-030-70051-5   ISBN 978-3-030-70049-2   (eBook)
https://doi.org/10.1007/978-3-030-70049-2

This Springer imprint is published by the registered company Springer Nature Switzerland AG
The registered company address is: Gewerbestrasse 11, 6330 Cham, Switzerland

Preface

This book was created over a period of more than 10 years and is based on the experience from classes taught as part of the renewable energy engineering program at the Oregon Institute of Technology. More than 500 program graduates, who work now in all aspects of renewable energy, have indirectly participated in the book creation through their insights, good and bad exams, assignments, and mostly passion to contribute to progress in energy generation and to protect the environment.

While the authors have collective experience in the energy field of close to 50 years in academia, industry, research labs, and humanitarian use of renewable energy, the educational approach when writing this book came mainly from direct experience in the classrooms and from the course notes. Over the years, the instructional approaches to teaching the topic were evolving until the right mix of the fundamental principles, technology description, device or plant designs, and environmental impact was established.

The book is an introduction to renewable energy, and it is deliberately written using easily understandable terminology and the right amount of information, without the need for upper-level mathematics or science. Every chapter provides practical examples, case studies, and simple exercises.

This book is primarily intended for engineering and science students, but also students from other disciplines with a general interest in renewable energy and curious high school students and professionals who need a quick introduction to renewable energy.

The authors hope that the book will inspire readers, who are first introduced to the subject, to develop passion and continue the study of renewable energy. The excitement of learning about solar and wind energy, hydroelectricity and biomass, ocean and geothermal energy, energy storage, and grid integration of renewables might be a step towards a better future for all.

Key Features of This Book
- Simple language
- Highly focused content
- A case study at the end of each chapter
- Concise, precise, and updated

Klamath Falls, OR, USA   Eklas Hossain
Wilsonville, OR, USA   Slobodan Petrovic

Acknowledgments

We would like to thank hundreds of renewable energy engineering students, who have inspired us, with their insights, creativity and passion, to put our experiences in teaching renewable energy into a book.

Contents

About the Authors

Eklas Hossain is an associate professor in the Department of Electrical Engineering and Renewable Energy and an associate researcher with the Oregon Renewable Energy Center (OREC) at the Oregon Institute of Technology (OIT), which is home to the only ABET-accredited BS and MS programs in renewable energy. He has been working in the area of distributed power systems and renewable energy integration for more than 10 years and has published many research papers and posters in this field. He is currently involved with several research projects on renewable energy and grid-tied microgrid systems at OIT. He received his PhD from the College of Engineering and Applied Science at the University of Wisconsin Milwaukee (UWM), his MS in Mechatronics and Robotics Engineering from the International Islamic University of Malaysia, and a BS in Electrical & Electronic Engineering from Khulna University of Engineering and Technology, Bangladesh. Dr. Hossain is a registered professional engineer (PE) in the state of Oregon and is also a certified energy manager (CEM) and renewable energy professional (REP). He is a senior member of the Association of Energy Engineers (AEE) and an associate editor for IEEE Access, IEEE Systems Journal, and IET Renewable Power Generation. His research interests include the modeling, analysis, design, and control of power electronic devices; energy storage systems; renewable energy sources; integration of distributed generation systems; microgrid and smart grid applications; robotics; and advanced control system.

Slobodan Petrovic is a full professor at the Oregon Institute of Technology (OIT), which is home to the only ABET-accredited BS and MS programs in renewable energy. Prior to OIT, he was a professor at the Arizona State University and held appointments at several companies as vice president and director of engineering. Dr. Petrovic received a BS in physical chemistry from the University of Belgrade, and his PhD in chemistry from the Technical University of Dresden. He has more than 30 years of experience in various areas of science and technology, is the author of the books *Battery Technology Crash Course: A Concise Introduction* and *Electrochemistry Crash Course for Engineers* (Springer 2020), has published more than 80 articles in international journals and conference proceedings, and has 35 pending or issued patents. Dr. Petrovic is a founder and president of the humanitarian organization Solar Hope, dedicated to providing solar energy to countries in Africa.

Renewable Energy Basics

Renewable Energy Definition

Renewable energy can be defined as energy that will not deplete naturally and can be extracted for an indefinite time. Renewable energy sources such as solar, wind, hydro, bioenergy, ocean energy, and geothermal are freely available from nature and do not harm the environment when converted to energy in the way fossil fuels or nuclear energy do. The carbon emission from utilizing renewable energy is minimal compared to energy conversion from fossil fuels or nuclear processes.

The sustainability definition includes a broader meaning and refers to energy generation that is not depletable on a human timescale, which does not lead to emission of greenhouse gases or other pollutants and which does not contribute to social injustice or geopolitical imbalance.

Energy Basics

Energy is defined as the ability to do useful work. While this may sound abstract and difficult to conceptualize for those new to science and engineering, it should be clear to all readers that energy is critical for sustaining life on Earth. It is also largely understood that energy may be converted from one form to another and that there are two basic expressions of energy, potential and kinetic, and that some energy is stored.

Examples of the stored form of energy are chemical energy in gasoline and gravitational energy of water ($E = mgh$). Other forms of energy include mechanical energy, thermal energy, which is kinetic energy of atoms and molecules, the energy of light (electromagnetic radiation, $E = h \times \nu$), electrical energy, magnetic energy, atomic energy, molecular energy, and the mass-energy interaction as expressed in Einstein's famous equation ($E = mc^2$).

© The Author(s), under exclusive license to Springer Nature
Switzerland AG 2021
E. Hossain, S. Petrovic, *Renewable Energy Crash Course*,
https://doi.org/10.1007/978-3-030-70049-2_1

Fundamentally, there are only three original sources of energy, solar nuclear fusion, heat from the core of the earth, composed almost equal parts of residual heat from the early stages of the planet and from nuclear processes in the Earth's crust, and energy from the rotation of the Earth and the moon. All energy on the Earth comes from these original sources of energy, which subsequently create primary or unprocessed energy. Examples of primary energy include fossil fuels (coal, oil, and natural gas) and renewable energy such as hydroelectric, solar, wind, biomass, ocean, and geothermal. The primary energy is then changed into effective energy that reaches the end-user in the form of electricity, heat, steam, gas, or hot water.

The Physical Laws of Energy

According to the First Law of Thermodynamics, energy is conserved, and it cannot be created or destroyed and can only be converted from one form to another. This also means that the total energy of the universe is constant. A system under observation may gain or lose energy, but the same amount must be lost or gained by the environment (or the surroundings). Energy is not created in the sun but converted from nuclear fusion energy into radiant energy.

A power plant, for example, a renewable energy power plant such as hydroelectric or solar, or a fossil fuel coal power plant, does not create energy—it only converts it from one form to another.

It should be understood by the reader that conservation of energy is different from *energy* conservation, which means using energy wisely.

Power

Power is mathematically the derivative of energy. It is the change in energy over time. The unit for power is Watt (W). For example, if 1 J of energy is given by a system in one second, power of 1 W is applied. Two systems may have an equal amount of energy stored, for example, in a battery, but if one system is able to deliver that energy faster than the other then, then that system is more powerful.

Another way to define power is to describe it as the production of energy. For instance, if a power plant operates at 1 GW of power, it produces 1 billion J of energy every second.

Energy Units and Conversions

Energy is the product of power and time and is most commonly expressed in practical systems in kilowatt-hours (kWh) instead of Joules, the unit of measurement used in physics. For example, if a power plant generates power of 1 MW (1×10^6 W) for 1 h, it produces 1 MWh (1×10^6 Wh) or energy.

Besides kWh, there are several other ways to express energy (Table 1.1).

Table 1.1 Energy units and conversions

Conversions	kJ	kcal	kWh	m³ gas	BTU	Therms
1000 J = 1 kJ	1	0.239	0.000278	0.000032	0.9478	0.0000095
1000 cal = 1 kcal	4.187	1	0.00116	0.00013	3.968	0.0000397
= 1 kWh	3600	860	1	0.113	3412	0.0341
1 m³ natural gas	31,736	7580	8.816	1	30,080	0.301
1 BTU	1.055	0.252	0.000293	0.000033	1	0.00001
Therms	105,480	25,200	29.3	3.326	99,976	1

There are also practical energy units based on oil and the fact that when oil is burned with oxygen, its chemical energy is converted to heat energy. One tonne of oil equivalent (toe) is simply the heat energy released in the complete combustion of 1000 kg of oil that produces on average 41.88 GJ.

Another unit used to measure the amount of energy is a barrel, which is 42 US gallons (35 Imperial or British gallons) or about 160 L. On average, there are 7.33 barrels in a tonne of oil. The energy content of 1 barrel is approximately 5.71 GJ, which is the unit called 1 barrel of oil equivalent (boe). Typically, it is expressed in million barrels daily (Mbd). The energy content of gasoline is about 44 GJ/tonne or 33 MJ/L.

There are also energy units based on coal. One tonne of coal equivalent (tce) is the amount of heat energy released when 1 metric ton of coal is burned. Although variations are significant between coal types, it is accepted that this value is equivalent to 28 GJ.

A British Thermal Unit (BTU) is the heat energy required to warm 1 pound of water by 1 °F. It equals 1055 J, while 1 therm is a large unit equal to 100,000 BTU.

One quad is a quadrillion BTUs equal to 1.055 EJ. Finally, 1 calorie (cal) is the heat energy needed to warm up 1 g of water by 1 °C and is equal to 4.19 J.

Energy Generation

The global energy consumption in 2019 was 580 EJ (580×10^{18} J) or 0.161 EWh (161×10^{12} kWh). A breakdown of the contributions from different technologies is shown in Table 1.2 for the United States and the world.

Renewables include hydroelectric, biomass, wind, solar photovoltaic, and geothermal. The data in the table reveals that fossil fuels combined contribute to over 84% of all energy produced in the world and about 70% of the US generation.

Fossil Fuel Reserves

Fossil fuels originate from organic matter that decomposed many hundreds of thousands or millions of years ago. Oil, coal, and natural gas used today to generate energy are the biological materials formed in the distant past under the conditions of heat and pressure. Because there are finite quantities of fossil fuels available in

Table 1.2 A breakdown in percentage of the energy generation technologies (sometimes called energy mix) for the United States and the world in 2019

Resource	% of U.S. Production, 2019 [1]	% of Global Production, 2019 [2]
Coal	11.3	27.04
Natural gas	32.1	24.23
Oil/ petroleum	36.7	33.06
Nuclear	8.5	4.27
Renewables	11.5	11.42

the Earth's crust, their future availability depends on the ratio of the rate of use and the number of reserves. While new fossil fuel material is still formed, the time scale of formation of millions of years means that we are now exploiting the reserves, and it is clear that the present rate of extraction will inevitably lead to depletion.

The quantity of the crude oil and natural gas reserves that can be recovered and exploited is often described using the so-called Hubbert Curve. It was established in 1956 by American geophysicist M. King Hubbert that the explorations of crude oil from a certain region first increase after discovery but eventually reach a peak after which the amount gradually decreases until depletion. The main conclusion from the Hubbert's historical resources model of crude oil discovery and exploitation is that there are finite reserves of crude oil in a certain region and, by extension, on Earth. The quantity of reserves or availability over time depends on the rate of extraction and follows roughly a bell-shaped curve, leading in the end to a complete depletion of reserves. For example, crude oil production in the United States peaked in the early 1970s while in the polar regions peaked around 2015. The exact shape of the curves is different for each region of the world; however, it is obvious that the peak in the global crude oil availability was in the first part of the twenty-first century, and the amount of crude oil available in the world is now on a sharp decline.

It is interesting to look at the availability of proven reserves and to relate it to current production (extraction) rates. The resulting number is the reserves-to-production (R/P) ratio or RPR, which is the ratio of the remaining amount of a nonrenewable resource to the rate of extraction or production. This ratio predicts how many more years the resource might last at a certain rate of extraction.

Consequently, it is generally understood that coal will remain a viable fossil fuel source for several more centuries, but oil and natural gas are expected to completely diminish in the next few decades at the most. For about two centuries, since oil was discovered in Titusville, PA, in 1859, the world's economically recoverable oil reserves have been nearly consumed, while it took millions of years for this resource to form from biomass.

The diminishing reserves of crude oil and natural gas, compounded by the harmful emissions from burning fossil fuels, have created the biggest turning point in the history of energy generation. The disadvantages of continued reliance on fossil fuels are obvious, making the case for the development of renewable energy technologies.

Renewable Energy Sources

The renewable energy sources and corresponding energy generation technologies include bioenergy, hydro energy, solar, wind, ocean, and geothermal.

Bioenergy is used in the form of wood, biomass, and crops (such as sugar cane) to produce heat and electricity. Wood and other bioenergy feedstock have the potential to be sustainable fuel sources if the rate of reforestation exceeds the rate of use. If managed properly, bioenergy can be exploited without a net increase in CO_2 emission. Bioenergy can also be converted to biofuels used for transportation. Although gaining global importance, bioenergy is currently a significant energy source only in developing countries.

Hydropower is the largest renewable energy contributor providing around 17% of the world's electricity in hydroelectric power plants. On a small scale, hydropower satisfies the sustainability criteria with no adverse environmental impact. However, on a large scale, massive dam installations result in major interruptions to the environment and can further lead to methane gas emissions through anaerobic decay of vegetation.

Solar energy can be utilized in two ways, to produce electricity in a photovoltaic reaction or heat in solar thermal processes. The source of solar energy is in the continuous nuclear fusion reactions between hydrogen isotope atoms inside the Sun that create temperatures on the surface of roughly 6000 °C. Only a small fraction of the energy of the Sun reaches the Earth.

Wind energy is an indirect manifestation of the energy of the Sun. It is one of the most important renewable energy technologies. The wind turbines can be installed on the land and offshore. A special case of wind energy utilization is harnessing the power of ocean waves.

The tidal energy is created by the gravitational pull of the moon on the world's oceans. The technology involves installing water turbines where the topography of shores amplifies tidal effects. The concerns about tidal energy are extremely high capital costs and the impact on the wildlife.

The origin of geothermal energy is in the Earth's internal heat, which comes from the radioactive decay of long-lived elements and the residual heat of formation. The most convenient way to take advantage of geothermal energy is to capture the underground water at high temperatures. The extracted hot water can be used for heating or steam can be used to generate electricity. Geothermal energy is renewable on a larger scale, but it can run out locally.

Case Study: Countries with 100% Renewable Energy Generation Energy is primarily required for electricity generation and transportation purposes and can be generated using different forms of renewable energy. This initial case study demonstrates how some countries with favorable resource distribution have accomplished electricity generation largely or completely based on renewable energy sources. The list of countries mostly or fully converted to renewables for power generation includes Denmark (69.4%), Brazil (75%), Austria (80%), Norway (98.5%), Costa Rica (99%), Paraguay (100%), and Iceland (100%).

Iceland, for example, has 25% of its electricity produced from geothermal energy and 75% from hydropower. Its geothermal energy is used to heat 90% of all homes.

Paraguay has 70% of all energy and 99% of its electricity generated from the hydroelectric power plants, Itaipú, Yacyretá, and Acaray. The remaining energy comes from different types of biomass.

It is estimated that more than 100 countries are in the position to achieve 100% conversion to renewable energy by the year 2050. This includes transition to carbon-free (no CO_2 emission) electricity generation and use of renewable energy sources for heating and cooling.

Exercise 1

1. What is meant by renewable energy?
2. What are the different sources of renewable energy?
3. Why are renewable energy sources preferred over fossil fuels?
4. What are the units of power and energy?
5. What is the shape of Hubbert curve?

References

1. https://www.eia.gov/energyexplained/us-energy-facts/
2. https://ourworldindata.org/energy

Hydroelectric Power

2

Introduction

The global installed hydroelectric power capacity at the time of writing of this text (in 2020) was 1308 GW and about 16% or 4306 TWh of world's electricity was generated in 2019 by the hydroelectric power plants. In the same year, China alone generated 27.24% of the global hydroelectricity, largely from the world's largest hydroelectric plant, the Three Gorges Dam with a generation capacity of 22.5 GW.

Some countries derive a significant portion of their electricity from hydroelectric power (often, terminology such as hydroelectric power, hydropower, or hydro is used interchangeably). Paraguay, for example, generates almost 100% of its electricity from hydropower because of its small population and vast water resources. Norway, another water-rich country, derives 95% of its power from hydroelectric sources, followed by Brazil, and Iceland, both at around 75%. Canada derives 61% and the US 52% of the total electricity demand from hydroelectric power. Brazil and Paraguay derive a significant portion of their hydropower from a single dam, the Itaipu, which has a rated capacity of 14 GW and ranks second among the largest hydroelectric power plants in the world.

The Water Cycle

Hydroelectric power is a form of indirect solar power. The heat from the solar energy reaching the Earth evaporates 513,000 km^3 of water every year (0.98 m^3/m^2). The solar energy from the sun is converted to thermal energy in water molecules, which rise into the atmosphere and eventually fall back down on the Earth's surface as precipitation. The lifting and falling of this enormous volume of water are called the water cycle. Overall, around 22% of the sunlight striking the Earth's surface is used to drive the water cycle. The water cycle process is schematically shown in Fig. 2.1.

© The Author(s), under exclusive license to Springer Nature Switzerland AG 2021
E. Hossain, S. Petrovic, *Renewable Energy Crash Course*,
https://doi.org/10.1007/978-3-030-70049-2_2

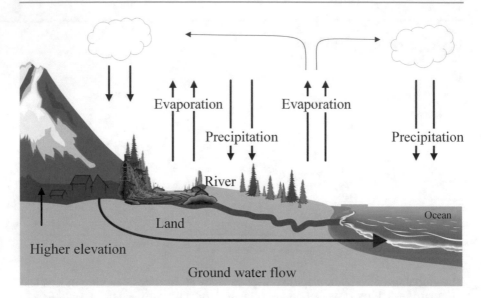

Fig. 2.1 Schematic of the water cycle

When water evaporates from the oceans (~87% of the overall evaporation), it rises into the atmosphere while gaining the potential energy coming from heat. About 96% of the sun's energy used in forming clouds is used for water evaporation and 4% for lifting water vapor against gravity. The water then falls back to the Earth's surface in the form of precipitation, and 78% of the rainfall occurs over the ocean. The remaining 22% of the precipitation falls over the land at different elevations, and the potential for energy extraction is realized from the flow of water from high elevations down toward lower elevations. During precipitation, i.e., when it rains, the gravitational potential energy is released mostly as kinetic energy and heat. A very small amount of precipitation that falls over the terrain higher than the sea level and then drives the flows of streams and rivers toward the seas is responsible for all hydroelectric potential, which is captured by building dams to concentrate potential energy.

Energy and Power of Water

When the sun's energy reaches the surface of the Earth, it heats oceans and lakes and evaporates water. The amount of energy required to evaporate 1 g of water is 2258 J. Another 100 J of the thermal energy of the sun is used to lift water vapor to the top of the troposphere at 10,000 m and acquire potential energy. The water then falls back on the Earth in the form of rain at different elevations and retains a portion of its gravitational potential energy. Some of this water flows in the form of rivers and streams from a higher elevation to a lower. The energy that can be obtained from the flowing water is calculated from the difference in potential energy at different elevations (Eq. 2.1):

$$E_P = mgh. \tag{2.1}$$

Here, m is the mass of water, h is the difference between elevations in the water flow path, and g is the gravitational acceleration constant of 9.81 m/s².

Electricity generators (i.e., turbines) placed in the pathway of this flowing water transform the kinetic energy of streams and rivers into electrical energy by slowing down the flow or utilizing the waterfall from a higher reservoir to a lower (Fig. 2.2). The most important characteristics of a site are the falling water "head," the height difference between the upper water level and lower (H_G) and the water flow. These characteristics determine the amount of energy provided by the falling water as it is directed through a pipe called a penstock. To calculate the available energy, the gross water "head" is reduced by the losses due to friction in the pipe and is then called a net "head," H_N.

The potential energy calculation can be demonstrated on the example of Grand Coulee Dam at the border of Oregon and Washington State in the US Pacific Northwest. The reservoir stores roughly nine billion cubic meters of water and has a height called the head of 170 m. The mass of water is determined by multiplying the volume by the density of water, 1000 kg/m³. To calculate the potential energy of this stored mass, the water height is averaged from top to bottom so that the average head height is 0.5 × 170 m or 85 m. Using Eq. (2.1), it is calculated that the potential energy behind the Grand Coulee is enormous.

$$E_{pot.} = 9 \times 10^9 \, \text{m}^3 \times 1000 \, \text{kg/m}^3 \times 9.8 \, \text{m/s}^2 \times 85 \, \text{m} = 7.5 \times 10^{15} \, \text{J} = 7.5 \, \text{PJ}.$$

From the expression for the potential energy of water (Eq. 2.1), the maximum power available for hydroelectric generation can now be calculated by considering the water flow, Q (in m³/s), and, after multiplying the volume with the density of water, the resulting mass flow (Eq. 2.2).

$$P_{max.} = 1000 \, \text{kg/m}^3 \times Q\left(\text{m}^3/\text{s}\right) \times g\left(\text{m/s}^2\right) \times h\left(\text{m}\right) \tag{2.2}$$

The most commonly used expression in the hydroelectric power industry (Eq. 2.3) contains correction factor, η, for the efficiency of converting kinetic power of the flowing water into electrical power in the turbines, the effective head (H_{eff}),

Fig. 2.2 Depiction of a dam in the water stream (left) and the water falling from upper reservoir into lower (right)

and the gravitational acceleration rounded to 10 m/s^2, giving the resulting power in kW.

$$P = \eta\, 10 Q H_{eff}\, (\text{kW}) \tag{2.3}$$

where the efficiency η is given as the ratio between the output electrical power and kinetic power, Q is in m^3/s, and effective head, H_{eff} in m.

Hoover Dam located at the Arizona-Nevada border near Las Vegas is 726 ft. tall (221 m) and contains 17 electric power generators capable of generating a total 2.2 GW of power. Around 330,000 gallons (1250 m^3/s) flow through these turbines every second. The effective head is estimated at 85% of the dam height and generation efficiency is 0.9 (i.e., 90%), giving the maximum power of 2.1 GW.

$$P = 0.9 \times 10 \times 1250 \times 187 = 2.1 \times 10^6 \ \text{kW} = 2.1 \text{GW}$$

In summary, the power output of a hydroelectric power plant depends on the dam head, or height, and the water flow through the dam. A short dam with a large flow can produce just as much power as a very tall dam with only a small flow.

Hydroelectric Power Plants

The first water-powered devices, the predecessors of modern hydroelectric power plants, date back to sixth century B.C. The earliest water mills appeared in the first or second century B.C. in the Middle East and spread later on to Scandinavia. In 1834, French engineer Benoit Fourneyron patented the first water turbine, characterized by the vertical axis, completely submerged blades, and over 80% efficiency. The first hydroelectric power station was opened in 1837.

Modern hydroelectric plants have a capacity from a few hundred watts to nearly 20,000 MW. They are classified based on the effective "head" of water, the type of turbine used, the capacity, the location, and the type of dam. The most common classification is based on the low, medium, and high "heads." There are three types of hydropower systems: impoundment involving dams (e.g., Hoover Dam, Grand Coulee), diversion or run-of-river systems (e.g., Niagara Falls), and pumped storage involving a two-way flow to a storage reservoir and return to a lower elevation for power generation. The main components of a conventional hydropower plant are a reservoir for storing water, a penstock for carrying water to the turbine, a turbine turned by the force of the water on its blades, and a generator driven by the turbine to produce electricity (Fig. 2.3).

Hydropower turbines are broadly classified as impulse turbines and reaction turbines. The turbine selection depends on the depth at which the turbine has to be installed, effective head, and water flow conditions. Impulse turbines are good for high water heads and low water flow. In these turbines, the water jet impinges on the curved, bucket-like blades resulting in kinetic energy that forces turbines to rotate. The velocity of impulse turbines is controlled by changing the diameter of the opening in the nozzle. There are three types of impulse turbines, Pelton, Crossflow, and Turgo turbines.

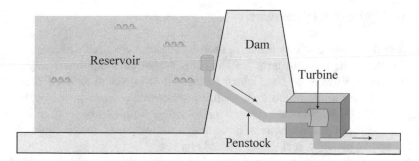

Fig. 2.3 Schematic representation of a hydroelectric power plant created by a dam

Fig. 2.4 Turbine selection based on water head and water flow rate

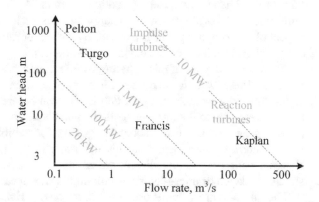

Reaction turbines are suitable for low water head and high-water flow. They utilize both the pressure and the velocity of the water. There are two types of reaction turbines, Francis turbine for medium water head and flow and Kaplan turbine for low water head and high water flow. Reaction turbines have flat but slightly angled or curved blades, which are pushed by the water flow. Turbine selection based on the water head and the flow rate is shown in Fig. 2.4.

Compared to other power generation methods, there are numerous benefits of hydropower such as low production cost, no emission of CO_2 and other harmful gasses, flood control (e.g., Three Gorges Dam), land irrigation, and waterway navigation. There are, however, also negative and unfavorable aspects of hydropower such as the potential catastrophic dam collapses, displacement of the population (the Three Gorges Dam project in China displaced one million people), destruction of cultures, and change in ecosystems. An environmentally harmful effect of hydropower dams is also the production of methane from biomass accumulated at the bottom of reservoirs. In the absence of a water reservoir formed by the dam, the biomass would decompose aerobically into CO_2 but when covered by water it decomposes anaerobically, producing methane gas, CH_4, which is 30 times more potent as a greenhouse gas than CO_2.

Hydropower water reservoirs behind the dams are also prone to silt accumulation, which over time reduces reservoir volume (and as a result the amount of stored energy) as well as the plant lifetime.

Micro-Hydroelectric Power

Small-scale hydroelectric is the term used to describe electricity generation on a small scale. In the U.S., the term "small-scale hydroelectric" applies to any project that produces 30 MW or less, which is enough to power around 6000 average American households. Micro hydro, on the other hand, refers to systems in the range 5–100 kW. Micro-hydro systems are relatively simple and easy to implement, e.g., AC generator may be nothing more than a car alternator, which could be connected up to a paddle wheel fed by a small creek.

A micro hydro-electrical plant powered by a water stream enables a continuous operation and is usually not interrupted by changes in weather, like solar photovoltaics or wind generators. The installation for micro-hydro is inexpensive compared with other small-scale renewable energy sources such as solar photovoltaics. The appeal of micro-hydro plants is particularly strong for remote locations, far from the electrical grid such as remote cabins and worksites.

There are two types of micro-hydropower generation systems, direct AC and DC systems. These systems share several components, an intake for water from the stream (called a weir), a channel, debris removal before the generator, and a pipe that carries water to the turbines (called a penstock). Finally, the powerhouse comprises a generator, an inverter, control circuitry for stabilizing power output, batteries for storing power, and transmission lines to the house or facility. Another channel, called tailrace, redirects water after the powerhouse back toward the stream. A sketch of micro-hydro installation is shown schematically in Fig. 2.5.

The micro-hydro systems use only a fraction of a river's flow and do not require dams to store water. Their implementation is straightforward and inexpensive compared to micro-wind and solar PV. However, the implementation is site-dependent. The river stream must be reliable with more than 3 m of the head (drop) and must have a steady water flow year-round to ensure optimum utilization. In addition, enough land must be available for the diversion weir, forebay tank, powerhouse, and transmission lines.

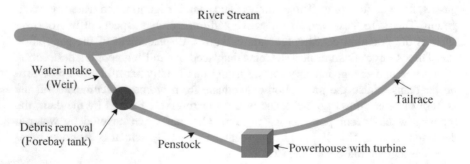

Fig. 2.5 Schematics of micro-hydro installation with main components, including a weir, channel, forebay tank, penstock, and a powerhouse

The electric power output for a micro-hydro system is calculated using Eq. (2.2) if the generator efficiency, η, the effective head height, H_{eff}, and the water flow rate are known. In one example, a weir diverts 50 gallons/s into a channel. After conversion, this is equivalent to 0.19 m³/s or (using the density of water of 1000 kg/m³) 190 kg/s. The head height, from the forebay tank to the generator through the penstock, is 4 m. The actual head height due to friction and turbulence losses within the penstock is assumed to be 80% of the original potential energy or 4 m × 0.8 = 3.2 m.

The generator is assumed to convert 75% of the kinetic power of the falling water into electric power, or the efficiency, η, is 0.75. Finally, the available power is:

$$P = 0.75 \times 190 \frac{\text{kg}}{\text{s}} \times 9.8 \frac{\text{m}}{s^2} \times 3.2\text{m} = 4.47\,\text{kW}$$

The system costs vary considerably depending on the required electric power, the proximity of the water supply to the consumer, and the type of system to be built, i.e., a direct AC system or a DC system. The initial costs are considerably less than equivalent micro-wind or solar installations but more expensive than comparable diesel or gas generators. An advantage of micro-hydro systems over diesel and gas generators is that there is no fuel cost as long as the water is flowing. The other advantages of micro-hydro systems include simplicity of construction and low maintenance costs, no gas emissions (compared to diesel and gas generators), long lifetimes of 20–30 years, and relative simplicity of the components needed.

There are two types of micro-hydro systems, direct AC systems and DC systems. Direct AC systems generate power from spinning turbine blades and power the loads (such as a household). As long as the water is flowing, the system produces power, which must be dissipated. Power in the excess of the loads can be used for a backup load, such as a water heater or air conditioning in the summer. It is important that the power demand does not exceed the capacity of the generator, even during the start-up surges (up to three times the power), to maintain the voltage and frequency within specifications. To ensure system availability and prevent collapses, a direct AC system must provide maximum power that a facility (i.e., a household) may require at any moment. Consequently, direct AC systems are usually oversized to account for power surges and require larger turbines at a higher cost.

DC systems operate differently because the power from a generator is used to charge the batteries, which then power the loads when needed. Since batteries produce DC electricity, an inverter is needed for converting to AC electricity necessary to power the loads. Because energy is stored in the batteries, the generator can be significantly smaller since it is generating power all the time, even when the loads do not need it. For example, a DC system equivalent to 2 kW AC system would have only 300 W generator. The smaller power rating allows for the use of smaller heads or lower water flow rates, so smaller hydro sites may be exploited. However, the system complexity increases because of the need for inverters and batteries. In general, DC systems are more complex than AC systems, but the system components, especially the generator, are much cheaper. DC systems are also more amenable for incorporation into hybrid systems with solar photovoltaics or diesel generators.

Case Study: The Itaipu Dam The Itaipu is the second-largest hydroelectric power facility in the world, behind the Three Gorges Dam in China. It was built on the river Parana, shared by Brazil and Paraguay. The construction of the Itaipu power plant began in 1971, and it had an expected lifetime of 200 years. From 1984, it has generated more than 2.7 thousand TWh of energy. The power plant is rated at 14 GW of power and has the capacity to generate 103 TWh of electric energy annually.

The dam is 7.9 km long and 196 m tall, and the reservoir occupies an area of 1350 km^2. The plant has 20 generators, each producing 700 MW of power. Together, they generate enough power to supply 10.8% of the energy demand in Brazil and 88.5% in Paraguay. Almost 90% of the energy generated by this plant goes to Brazil. Interestingly, Brazil uses 60 Hz frequency in their electric grid, while Paraguay uses 50 Hz frequency in their grid, while both receive power from the Itaipu hydroelectric plant. Half of the generating units of the plant generate power at 50 Hz and half of them generate power at 60 Hz. On the Brazil side, a converting station then converts the 50 Hz power into 60 Hz.

The Itaipu hydroelectric powerplant is a remarkable technological accomplishment and a major contributor to the world's total renewable energy.

Exercise 2

1. What is an upper absolute upper limit to the Earth's hydro capacity expressed in TWh if the total precipitation over the land area is 10^{17} kg per year and the mean height of land above the sea level is 840 m? ($g = 9.81$ m/s^2) Express with the correct number of significant figures [228.9 TWh].
2. Calculate the electrical power in kW delivered from a micro-hydro plant on a mountain stream with an effective head of 40 m and a water flow of 12,000 L/min ($g = 9.81$ m/s^2). The generator efficiency is 80%. Express the result with two significant figures [62.78 kW].
3. Apart from micro-hydro systems elaborated in this chapter, find out more about different types of hydroelectric power plants such as large hydro, medium hydro, small hydro, mini-hydro, and pico-hydro systems. Prepare short notes about them.
4. Analyze the differences between hydroelectric powerplants and pumped hydro storage.
5. What are the differences between impulse turbines and reaction turbines?

Wind Power

3

Introduction

The appeal of electricity generation from wind power has its foundations in the exceptional resource potential and great power density. As with a few other globally available technologies, wind power, if fully exploited, could completely satisfy the demands of the world for energy. Sharing similar characteristics with other renewable energy sources such as energy generation from wind power is relatively unpredictable and unreliable. The power density of wind varies from low, in case of a mild breeze to very high, in case of a hurricane. Wind is also dependent on atmospheric conditions and weather patterns. While the availability of wind is not completely random, the wind resource and power are only partially predictable. Furthermore, only a small part of the global wind potential is economically recoverable.

The modern use of wind power is based on turbines that produce electricity. However, historically, the wind was used in mechanical devices for milling grain and for pumping water. The earliest uses were reported over 3000 years ago and could after that be tracked to the twelfth century in Europe. In the seventeenth and eighteenth centuries, windmills were used in parts of Europe, and in the nineteenth century, they were already widespread in North America. In 1900, a wind turbine was used to electrolyze water and produce hydrogen for gaslighting. The development of wind technology for electricity production took place in the late twentieth and the twenty-first centuries.

Wind energy potential worldwide is estimated at over 53,000 TWh/year and could provide 12% of global electricity needs in 2020. The exploitation of wind power depends on the location and on the wind speed, which is typically measured using a Beaufort scale. Typically, winds below 3 m/s and above 25 m/s cannot be used to generate power.

© The Author(s), under exclusive license to Springer Nature
Switzerland AG 2021
E. Hossain, S. Petrovic, *Renewable Energy Crash Course*,
https://doi.org/10.1007/978-3-030-70049-2_3

Fig. 3.1 Sketch of the global winds

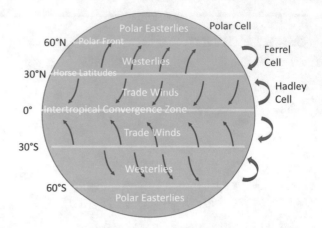

Origins of Wind Energy

From the solar radiation of 5.4 million EJ/year, only a small portion, a 11.7 thousand EJ/year, is driving the global winds, waves, and currents. Global winds form as a result of temperature differences and the rotation of the Earth. When the evaporation takes place on the equator, it creates a low-pressure area close to the ground, attracting winds from the North and South. Around 30° latitude on both hemispheres, the Coriolis force from the rotation of the Earth prevents the air from moving further. At this latitude, there is a high-pressure area, as the air begins sinking again. This creates a so-called Hudley cell of circulating air. At the Poles, there is a high pressure due to the cooling of the air and, as a result, Ferrel and a Polar cell are formed. This air circulation because of pressure differences, together with the rotation of the Earth, creates the global winds. These are named Trade winds, Westerlies, and Easterlies and shown in Fig. 3.1.

In addition to the global winds, there are local winds. Sea breezes are generated in coastal areas due to different heat capacities between the sea and the land. They blow from sea to land during the day and reverse at night. Mountain winds are created when cool air in a valley warms up in the morning and, as it becomes lighter, begins to rise. At night, the cool air from the mountain falls.

Energy in the Wind

The wind is the movement of air mass with kinetic energy calculated using Eq. (3.1). To calculate the kinetic energy in Joules (kgm²/s²), the mass is expressed in kilograms and the speed in m/s.

$$E_K = \frac{1}{2}mv^2.$$ (3.1)

The power or the flow of energy is then calculated in J/s by taking the time derivative of the kinetic energy with respect to wind mass at a constant velocity (Eq. 3.2).

$$P = \frac{d}{dt}\left(\frac{1}{2}mv^2\right) = \frac{1}{2}v^2\frac{dm}{dt}. \tag{3.2}$$

The mass can be expressed as a product of density and volume ($m = \rho V$), and then the volume of air can be substituted with the product of the surface area captured by the spinning turbine blades and wind velocity $\left(\frac{dm}{dt} = \rho A v\right)$. This means that the mass of air, with a density of ρ, flows through an area A with a speed of v. Finally, the power in the wind is expressed as a simple equation:

$$P = \frac{1}{2}\rho A v^3 \tag{3.3}$$

It can be further observed that the power in wind increases exponentially with speed by a factor of three, e.g., if the speed of wind doubles, the wind power goes up by a factor of 8. The density also affects the power, i.e., cooler air is denser than warmer air and therefore carries more power for a given speed. The temperature plays a role as well because the wind density depends on temperature (Table 3.1).

Using Eq. (3.3) and applying values for the air density, P/A from Table 3.1, it is now possible to plot the wind power density as a function of the wind speed to show that wind power increases with increasing wind speed and with lower temperature (Fig. 3.2).

The power density increases steeply due to the third power of speed term in Eq. 3.3. The slight variations in wind power density with temperature are the result of changes in air density. Another important conclusion from the graph is that the density of wind power even at moderate wind speeds is significantly higher than 1 kW/m^2, which is the maximum density of solar radiation on earth surface.

Power and Energy from Wind Turbines

For practical purposes, it is important to define the range of wind speeds useful for the operation of wind turbines. Relating to our experience, it is possible to say that a breeze can be from 2 to 5 m/s, 16 m/s might be considered a strong wind, and

Table 3.1 Air density as a function of temperature

Air temperature	Air density, ρ in kg/m^3
−10 °C, 14 °F	1.32
−0 °C, 32 °F	1.28
10 °C, 50 °F	1.23
20 °C, 68 °F	1.19
30 °C, 86 °F	1.15

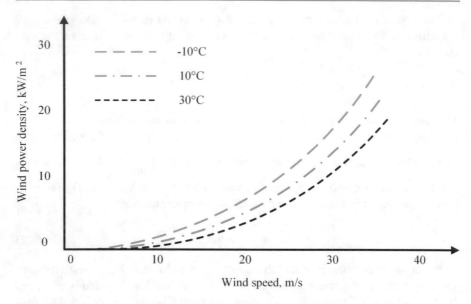

Fig. 3.2   Wind power as a function of wind speed for three different temperatures

35 m/s a hurricane. It is worth mentioning that hurricane winds are not used to produce power because they can cause significant structural damage to the spinning turbines. Most wind turbines are designed to start spinning at wind speeds between 3 and 5 m/s, which is called the cut-in wind speed, and programmed to stop spinning at a cut-out wind speed of 25 m/s.

A turbine converts the kinetic energy of air that passes through the cross-section swept by the rotor blades into rotational energy of the rotor, which in turn drives the generator and produces electrical energy. However, it is not possible to capture all this energy. The wind has certain energy when it enters the turbine where its speed is slowed down by the blades while a portion of its energy is transferred to the blades. The turbine slows down the airflow, but it never completely stops it. As a result, the air exits the turbine at a lower speed and with less power. The difference that was captured by the blades is converted into electricity. The amount of power that wind can transfer to the rotor depends on the density of air, the wind speed, and the rotor area.

The power extracted or captured by the turbine is the difference between the wind power before and after exiting the turbine. By rearranging Eq. (3.3), the following expression is obtained for the power captured by the turbine, P_T.

$$P_T = \frac{1}{2}\frac{dm}{dt}\left(v_1^2 - v_2^2\right) \tag{3.4}$$

In this equation, *dm/dt* is the mass flow, and the expression has been derived from the air density, cross section, and wind speed used in the Eq. (3.3). It should also be obvious that velocity v_1 before entering the turbine is higher than v_2 after the turbine.

Since the mass flow of air is not a useful or easily measurable quantity, another expression involves average wind speed through the turbine and cross-sectional area covered by the blades (Eq. 3.5).

$$P_T = \frac{1}{4} \rho A \left(v_1 + v_2 \right) \left(v_1^2 - v_2^2 \right) \tag{3.5}$$

The equation assumes that the actual wind speed at the blades is an average between the entrance and the exit speed and provides a good estimate of the power a wind turbine can generate. A more precise meaning can now be assigned to the area, A, and it is defined as the area circumscribed by the blades of the turbine. If a turbine blade is r meters long measured from a rotor, then the area of a circle with a radius of r is $A = \pi r^2$.

A figure of merit assigned to wind turbines is the power coefficient, C_P, which is the ratio of the power generated by a turbine, P_T, to the power of the incoming wind, P_I.

$$C_P = \frac{P_T}{P_I}. \tag{3.6}$$

After combining Eqs. (3.3) and (3.6), it can be shown that the power coefficient of a wind turbine is a function of the initial and final speeds of the wind:

$$C_P = \frac{\left(v_1 + v_2 \right) \left(v_1^2 - v_2^2 \right)}{2v_1^3}. \tag{3.7}$$

Equation 3.7 also shows that the power coefficient of a turbine is independent of the density of air and the cross-sectional area or the size of the blades. Instead, C_P is only a function of the turbine's ability to slow down the wind from v_1 to v_2. Further mathematical analysis of the expression in Eq. (3.7) leads to a conclusion that the maximum power coefficient for any turbine is 0.593. The coefficient is called the Betz limit or the Betz law after the German physicist Albert Betz who formulated it in 1919 and published in his book "Wind Energie" (note German spelling) in 1926. The Betz Law indicates that a wind turbine can convert a maximum of 59.3% or 16/27 of the power in the wind into mechanical power.

The maximum power coefficient is obtained when a turbine slows the wind speed to roughly one-third of the initial speed. The Betz limit is not some mysterious property, but similar to the Carnot efficiency for heat engines or the Gibbs Free Energy for fuel cells, it sets a limit to how much of one form of energy can be converted into another. The Betz law is often shown in a different form, as a ratio of the power extracted from the wind and the power of the undisturbed wind (Eq. 3.8).

$$\frac{P}{P_0} = \frac{1}{2} \times \left[1 - \left(\frac{v_2}{v_1} \right)^2 \left(1 + \left(\frac{v_2}{v_1} \right) \right) \right] \tag{3.8}$$

After establishing the dependence of the wind power upon the wind speed (Fig. 3.1 and Eq. 3.3), the electrical power curve for a wind turbine can be plotted as a function of the wind speed (Fig. 3.3).

There are several regions in this plot. First, the power output is zero and the turbine blades are locked until the "cut-in" wind speed is reached, with sufficient power to generate electricity. Once the wind speed exceeds the "cut-in" speed, the turbine blades are allowed to spin, and the turbine generates power output increasing with the third power of the wind speed. As the wind speed increases, the turbine reaches its nominal or rated power output. For example, a turbine rated at 1.2 MW will produce 1.2 MW only when the nominal or rated wind speed is reached. Above the nominal wind speed, the turbine rotation must be limited, as the centrifugal forces exceed specified values. As a result, there is no increase in power output for higher wind speeds than the nominal. This is accomplished by intentionally limiting the pitch angle of the turbine blades and slowing down the rotation of the blades. If wind speeds exceed the maximum value for a specific turbine, the pitch of the blades is adjusted away from the wind to stop the rotation of the blades. This is done to prevent damage to the turbine If the wind speed is too high. The typical cut-in wind speeds are 2.5–4.5 m/s, the nominal from 6 to 10 m/s, and the cut-out speed is from 20 to 30 m/s.

The energy produced by a wind turbine depends on the wind power curve (Fig. 3.2) and a wind speed frequency distribution for a particular site or the number of hours the wind blows at each speed (Fig. 3.4). The energy for a particular wind speed can then be calculated by multiplying the extracted power from a wind turbine by the number of hours. The total energy for the turbine is obtained by adding

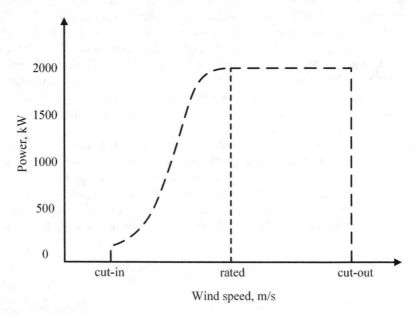

Fig. 3.3 Power output versus wind speed for a hypothetical wind turbine

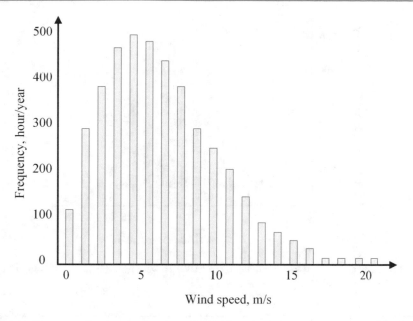

Fig. 3.4 Histogram showing the frequency distribution of wind speeds in number of hours per year for a hypothetical turbine site

all energies at different wind speeds and applying a correction factor that includes transmission losses and turbine availability, which depends on the reliability and maintenance issues (usually > 90%). An estimate of the total electricity production for a particular site can be obtained using Eq. (3.9).

$$E[kWh] = K \times v_m^3 \times A \times T \tag{3.9}$$

where K equals 3.2 is a factor based on typical turbine performance, v_m is the annual mean wind speed in m/s, A is the swept area of the rotor in m², and T is the number of turbines in the wind farm.

With knowledge of electricity produced for each wind speed and their frequency distribution, it is now possible to construct a predictive histogram of the electricity production for each wind speed.

By comparing the histograms in Figs. 3.4 and 3.5, it is observed that the distribution profiles don't match, i.e., the maximum of the frequency distribution is at a wind speed of 4 m/s, while the maximum energy production is for the wind speed of 10 m/s. Although this is a hypothetical site, an actual site would likely show the same trend because the generated power is higher for the stronger winds, while milder winds are more common. It can also be observed from the histogram in Fig. 3.4 that no electricity is generated for wind speeds below approximately 3 m/s.

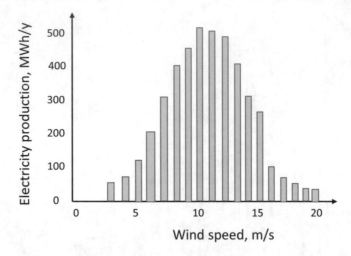

Fig. 3.5 Histogram of electricity production in MWh per year for each wind speed for a hypothetical site

Turbine Properties

The power extracted by a wind turbine corresponds to the square of the turbine radius, inferring that doubling the radius or the length of the blades increases the power by a factor of 4. By combining Eqs. (3.3) and (3.6), an expression for the electrical power generated by a turbine is derived, showing dependence on the surface area swept by the rotor, πr^2, and the power coefficient, C_p.

$$P = \frac{1}{2}\rho\pi r^2 v^3 C_p. \tag{3.10}$$

The radius of the wind turbine can now be defined (Eq. 3.11)

$$r = \sqrt{\frac{2P}{\rho\pi v^3 C_p}}. \tag{3.11}$$

It is revealed that the radius of a wind turbine can be calculated using the desired output power, the air density at the location, the wind speed at the rated power, and the coefficient of performance of a turbine, which is known for any specific turbine type.

The electrical power that can be generated from a wind turbine depends on both the height above the ground and the altitude above the sea level. Both conditions result in lighter air, which produces higher wind speeds. The height of the rotor from ground level is known as the hub height of a wind turbine, and it can range from 25 to 100 m. The relationship between hub height and increase in turbine power is shown in Fig. 3.6.

The change in wind speed with altitude is known as wind shear. This relationship, referred to as Power Law, can be expressed as,

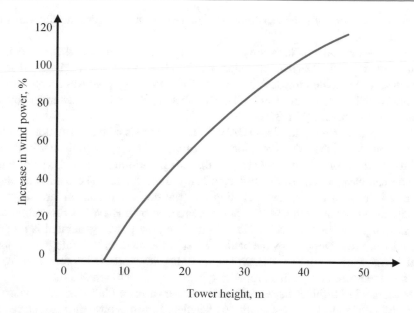

Fig. 3.6 Relation between hub (i.e., tower) height and extracted wind power. Rated power is assumed for the tower height of 10 m and the heights above that result in a higher power

$$\frac{v_2}{v_1} = \left(\frac{H_2}{H_1}\right)^{\alpha}.$$

(3.12)

Here, v_1 is the measured wind speed at a known altitude H_1, and v_2 is the estimated wind speed at another altitude H_2. Exponent α is the wind shear factor, which varies with the location, temperature, altitude, time of day, season, and atmospheric stability. Higher values of α indicate larger changes in the wind speed with vertical elevation. The value of α ranges from 0 to 1. For stable or neutral atmospheric conditions, α is assumed to be 1/7 or 0.143.

Turbine Types

Wind turbines are classified as vertical axis turbines and horizontal axis turbines. Vertical axis wind turbines (VAWTs) have the rotational axis perpendicular to the surface where they are installed and the direction of wind flow. These are the oldest type of windmills, with some evidence pointing to their use over a 1,000 years ago.

VAWTs are advantageous over HAWTs, as they may extract wind blowing in any direction without re-orienting the rotor with changes in wind direction. This makes them suitable for locations with quickly shifting winds. Another advantage of VAWTs is in a simpler structure that is easier to assemble. However, several disadvantages have restricted the use of VAWTs to specialized applications, limiting widespread implementation as in the case of horizontal axis wind turbines (HAWTs).

These include poor efficiencies compared to HAWTs and larger material stress on their blades.

The only vertical axis turbine which has ever been manufactured commercially at any volume is the Darrieus machine, named after the French engineer Georges Darrieus, who patented the design in 1931. This type of turbine was manufactured in the United States until 1997. The Darrieus machine is characterized by its C-shaped rotor blades (Fig. 3.7).

Horizontal axis wind turbines (HAWTs) have the axis of blade rotation parallel to the ground, i.e., the axis of rotation is horizontal with respect to the earth surface and to the direction of wind flow (Fig. 3.8). This predominantly used type of wind turbine can have one, two, or most commonly three blades. The choice of blade number is a result of analysis including material costs, mechanical stresses, efficiency, and visual smoothness. The power generated by a HAWT is proportional to the second power of the turbine blade length, i.e., the power generated is proportional to the area swept out by the blades. Most of the modern technology uses horizontal axis wind turbines (or HAWTs), and all grid-connected commercial wind turbines today are built with a propeller-type rotor on a horizontal axis.

Modern wind turbines have an uneven number of rotor blades mainly to ensure the stability of a turbine. Most are three-bladed designs that have the rotor placed on the windy side of the tower. This is called a classical Danish concept and is considered a standard to compare other concepts. Because the wind direction always changes, the turbines have a yaw mechanism used to turn the turbine rotor against the wind.

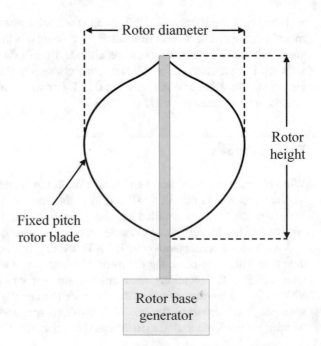

Fig. 3.7 Sketch of a vertical axis wind turbine after Darrieus

Fig. 3.8 Sketch of a
horizontal axis wind
turbine

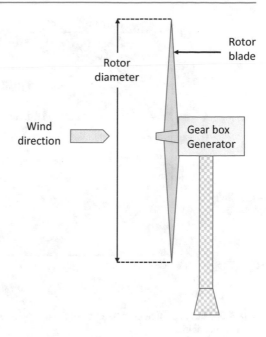

Two-bladed and one-bladed wind turbine designs have the advantage of a lower cost and weight, but they require higher rotational speed to generate the same energy output as the three-bladed turbines. This leads to higher noise and visual intrusion. A comparison of the turbine efficiency or power coefficient, CP, based on the number of blades, as a function of the ratio of the tip speed to the wind speed, is shown in Fig. 3.9.

It is also worth mentioning that wind turbine spacing is an important issue for wind farms used for utility-scale power generation. A wind turbine cannot be placed in the vicinity of another wind turbine, and, in general, a minimum separation of 7 rotor diameters between adjacent wind turbines is necessary. For example, for a typical turbine of 1000 kW, a tower height of 60–80 m and a rotor diameter of 54 m must have a minimum distance of around 350 m to the next turbine on all sides.

The turbine sizes, reported using the power produced and the diameter of the rotor blades, have evolved over the years. In the 1980s, the turbines varied in size from 100 to 300 kW, with the rotor blade diameter 15–30 m. The tower was typically made of steel, the generator was induction type, and the blades were made of fiberglass. In the 1990s, the power generation capacity increased to 750 kW, which became a standard size for a while, with a rotor blade diameter 30–50 m. A variable speed and new airfoil, i.e., blade technology, was introduced in this period. In the 2000s, the size increased to several MW, the tower became more flexible, the blade materials further improved, and low speed, direct-drive generators were introduced. The largest wind turbine at the time of writing the text is 12 MW near Rotterdam in Holland, having one blade of 107 m. The turbine produces enough power for 12,000

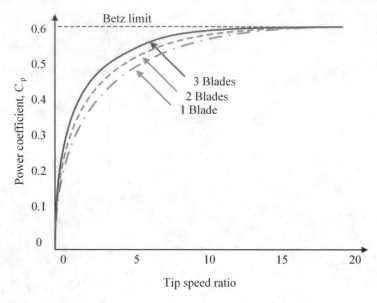

Fig. 3.9 Power coefficient as function of the tip speed ratio to wind speed for turbines with different number of blades

households in Europe. The wind turbine technology also changed from inland turbines, to offshore, and then to deep water wind platforms.

System Components

The main components of a wind turbine are the rotor blades on a low-speed shaft, connected to a gearbox that transfers the power to a generator on a high-speed gearbox. The rotational speed increases from about 30–60 rotations/min in a low-speed gearbox to between 1000 and 1800 in a high-speed gearbox. The generator is usually an induction generator that produces 60 Hz AC electricity. The gearbox is one of the most important wind turbine components, and alternate designs have been developed to make "direct-drive" generators that operate at lower rotational speeds, without the need for gearboxes.

A controller starts up the turbine at cut-in wind speed and shuts down at cut-out speed. A wind vane measures wind direction and communicates with the yaw drive to orient the turbine rotor towards the wind.

The breaks prevent the blades from spinning when the turbine is not generating power, at low wind speed, and also when the wind speed exceeds the cut-out point. The breaks are applied to bring the blades to a complete stop but only after turning them out of the wind.

Wind Power Generation Overview

As with any energy technology, there are advantages and disadvantages of wind power generation that must be considered. Wind power does not require a large land area. For example, one of the largest wind projects in the Western United States, the Stateline Energy Center, deep in the Columbia Gorge, on the border of Oregon and Washington, requires around 200,000 ha (75 miles^2) and generates about 300 MW of power, sufficient to power about 60,000 homes. The footprint of a single turbine is on average 1.2 ha/1 MW of delivered power, while the overall area around the turbines must be left open and it is 20 times larger. However, this area may be repurposed for activities such as farming. In comparison, power plants that burn coal or nuclear power plants using uranium may have a lesser footprint, but the mining operations required to supply fuels to these plants destroy enormous areas, e.g., the whole mountain tops in Kentucky, West Virginia, and Tennessee.

In general, wind power plants offer an excellent investment opportunity and a large return. A steady increase in wind power global capacity and a growing popularity of wind energy come as a result of the combination of favorable performance, good environmental record, and low cost.

However, there are some factual and perceived disadvantages of wind power, which must be addressed and mitigated through advanced design and implementation of wind power systems.

The primary concern with wind power is that it is intermittent and not always available when electricity is needed. If wind energy is produced at time when there is no demand, e.g., at night, or if its output varies during the day, integrating it with the electrical grid becomes more challenging. Because the wind is blowing intermittently and at varying speeds, the power cannot be dispatched like a conventional energy source, and wind turbines deliver a variable level of power. The main concern for the opponents of wind power is related to reliability and timing mismatch between the generation and demand. The argument often used is that electrical grid operators do not like the changing output of the generation coming from the wind turbines. In reality, the complete power system and its electrical grid are designed to deal with load and generation variations. Adding a variable wind generation to the system makes it more complex, but grid operators have good experience in handling the variations with the help of fairly accurate long-term models for wind generation. However, adding wind generation to a system increases the cost of balancing, which could be about 10% of the overall generation cost and depends on the resource mix, geographical considerations, and the amount of wind energy penetration. In general, more wind generating capacity simplifies balancing the variability. Ultimately, the best solution for an intermittent generation from the wind is to add batteries, or water electrolysis with hydrogen fuel cells, or another energy storage method, and use it at the time of the demand.

Environmental Impact of Wind Generation

The positive environmental impacts of wind turbines include emission-free opera-tion, no direct connection to mining or extracting natural resources, and a very favorable cost compared to coal and natural gas power plants. The lifetime analysis shows that a wind turbine generates about 80 times the energy required to build it. In comparison, biofuels generate 1.5 times as much energy as is used to create them, and gasoline only about 0.8 times the energy used to extract it.

One of the issues often cited against wind power is the noise coming from the turbines that may impact people living in the vicinity. Upon analysis, it becomes apparent that the noise generated by the turbines is significantly less, for the same amount of power than that generated by the steam turbines used in coal, natural gas, or nuclear power generation. Furthermore, the turbine designs are undergoing con-stant improvement through better engineering to reduce the noise. Over the years, the noise affecting people living in the proximity of the turbines has significantly decreased. There are two sources of noise: the aerodynamic noise from the spinning blades and the mechanical noise from the metal components moving against each other in the gearbox, the shafts, and in the generator. Older types of wind turbines were producing mechanical noise that could be heard up to distances of 200 m, but most modern designs have eliminated this problem. Aerodynamic noise from the spinning turbine blades increases with the fifth power of the blade speed. Improvement in the blade design has reduced the aerodynamic noise as well and, at the same time, led to increased turbine efficiency. Typical noise levels from a wind turbine are shown in Fig. 3.10 and compared to some commonly known noise levels.

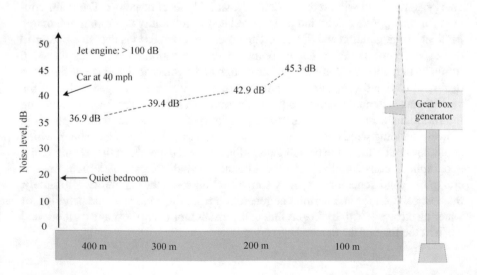

Fig. 3.10 Average noise level fading as a function of a distance from a turbine for a hypothetical wind turbine site

The presence of wind turbines and farms changes the visual appearance of an area and sometimes obstructs views. The large wind farms, comprising many small turbines, such as Altamont Pass and Palm Springs in California, built during the 1970s and 1980s, are sometimes described as a visual disturbance of the scenery. Since then, it has been widely accepted that the visual impact may be mitigated by design, for example, by building farms of uniformly spaced turbines of the same size, color, and shape and creating a more attractive panorama than a farm that is not organized by aesthetic design considerations.

Shadow flicker is another perceived drawback of wind turbines. It is a phenomenon caused by light disruption as the blades of a spinning turbine periodically block out the sun. This is generally only a problem for people living close to wind farms. Shadow flicker may be reduced by advanced blade design and modeling and by more optimal positioning of the turbines.

The arguments against wind turbines because of bird and bat collision incidents are widespread. The reason for collisions is that birds and bats cannot see the rapidly spinning blades of a turbine. However, the problem is usually associated with smaller, older models of turbines, those that generate 200 kW of power or less, because of their rapidly spinning blades that appear transparent to the birds. The newer, larger generators, producing megawatts of power, have slower angular speeds and their blades are much easier to see. Coloring at least one blade of a turbine with dark color can further improve visibility.

Incidents involving birds are site-specific, and some studies stating a large number of incidents have not been confirmed at other large wind farms. It is also informative to consider the comparative analysis of the impact of other energy-related sites on birds' deaths. For example, the Exxon Valdez oil spill killed an estimated 500,000 birds. A comprehensive analysis shows that the impact of wind turbines on bird deaths is relatively small when compared to the number of birds killed by cars, those hitting windows, or in other types of incidents; and overall, it appears that the wind turbines are responsible for no more than 0.01% of all bird deaths. As the number of turbines grows, bird accidents may increase, but the bird deaths measured per GW of generated power should fall significantly as larger turbines become prevalent.

Electromagnetic interference is not an environmental issue, but nonetheless, there are concerns with the interference of electromagnetic signals by the large metallic blades of spinning turbines. Metal blades of 100 ft. in length or more can significantly affect the transmission of radio, television, and microwave relay signals. However, modern turbine designs utilize lighter materials such as carbon-reinforced fibers or fiberglass that do not interfere with the transmission of electromagnetic signals.

Case Study 3: Location of Wind Turbines The site selection for wind turbines is a vital consideration, and wind turbines can be located onshore, along coastal areas, or offshore, away from the coastline. Wind resource potential is very high along the coastal areas, where wind moves in an uninhibited way across the surface of the ocean. In addition, the large temperature differences between the ocean and the shore create strong, predictable winds.

Examples of wind power installations along coastlines include the Mid-Atlantic and Northeastern coasts, the Great Lakes coasts, and the Southeast coast of Texas. Offshore wind farms constituted about 4.5% of the global cumulative wind power installations in 2019. Although more expensive than onshore installations, the offshore turbines are larger and can generate more power. The Hornsea Project One, located off the Yorkshire coast in the North Sea, rated at 1.2 GW, is the largest offshore wind farm in the world. There are plans for the construction of two more projects rated at 1.4 and 2.4 GW.

The Netherlands is famous for its windmills. The country relies heavily on wind power, and all electric trains, consuming 1200 GWh annually, are solely powered by the wind energy. Windpark Noordoostpolder, a 429 MW wind park in Flevoland, Netherlands, along the Ijsselmeer lake, comprises three subprojects, two onshore wind farms, and one offshore wind farm, with a combined capacity of 1400 GWh, sufficient to power 400,000 households. The park has 86 turbines in total, 38 onshore and 48 offshore or near-shore.

Exercise 3

1. Make a comparative analysis between horizontal axis and vertical axis wind turbines.
2. What is the power coefficient of a wind turbine if the initial and final wind speeds are 20 m/s and 8 m/s? [0.59].
3. What is the power that a wind turbine with a blade radius of 18 m receives from the wind that has a speed of 25 miles/h and operates at $30°C$ temperature? [817 kW].
4. Describe the advantages and disadvantages of onshore and offshore wind turbines.
5. At a certain place, if the wind speed at an elevation of 30 m is 2.6 m/s, what is the wind speed at a height of 60 m from the Power Law equation? Consider $\alpha = 0.2$. [2.987 m/s].
6. If the wind speed is doubled, what is the increase in the power extracted by a wind turbine? Show mathematically. [The power increases by 8 times].
7. What are the feasibility considerations you need to study to design a 5 MW wind farm? What kind of hub height, turbine size, and area would you choose for the farm and why? Discuss your views regarding the design of a wind farm.

Ocean Power

4

Introduction

Technologies for deriving electrical power from the ocean are based on wave energy, tidal energy, and ocean thermal energy. The wave energy is used by converting the kinetic energy of the moving waves into mechanical energy powering water turbines. Tidal energy is harnessed by collecting water in reservoirs behind dams and powering the water turbines similar to hydroelectric power plants. Tidal power generation requires large tidal differences and is limited to such coastal locations. Power plants can also be built to utilize the temperature difference between the warmer surface water and the colder deep ocean water. Ocean thermal energy conversion is limited to tropical regions.

Wave Power

Waves are created when the wind passes across the surface of the water of oceans, lakes, or rivers and cause it to move. The waves don't actually move water and instead, they transmit energy. The origin of this energy is the solar radiation on the Earth and a small portion of 5.4 million EJ/year, which is first converted to wind and an even smaller portion to waves. This makes the wave energy a stored and concentrated form of solar energy.

Besides wind-driven waves, there are also ocean waves caused by underwater disturbances, earthquakes, and volcanic eruptions. These waves are called tsunamis and are not used for power generation.

The wind-driven waves are common across the open oceans and along the coasts. They are typically created in deeper water, further from the shores, and then increase in amplitude and the amount of energy they carry as they travel toward the shores.

© The Author(s), under exclusive license to Springer Nature
Switzerland AG 2021
E. Hossain, S. Petrovic, *Renewable Energy Crash Course*,
https://doi.org/10.1007/978-3-030-70049-2_4

Worldwide Potential for Wave Power

Among the three forms of ocean power, waves have the highest global resource potential estimated at 2 TW (2000 TWh/year), which is at least 7% of the total global electricity demand.

The resource potential for wave power depends on the energy density of the ocean waves and the length of the coastline. These sites are located in oceans at the continental shelf drop-off lines. The continental shelf borders are underwater, slopped extensions of the continents that end in a drop-off or shelf break, deep on the ocean floor. The waves formed in these regions have a low amplitude that gets amplified as they travel toward the coastlines, where they become high amplitude waves that can be used for power generation.

As expected, the wave formation depends on the winds. Trade winds and the Westerlies create the highest power waves along the western continent coasts and between the latitudes of 40° and 60°. However, other winds in combination with coast characteristics can generate high overall power on the eastern shores of the continents as well.

There are numerous coastlines in length from 400 km to more than 1000 km and with wave power up to 100 kW/m. Several of such coastal locations alone have the potential to provide enough energy from deep-water waves to satisfy 10% of the total global energy demand. These best wave power generation coastal regions include the SW of Ireland in the North Atlantic Ocean, Antarctic Ocean, and the intersection of the South Pacific Ocean and the South Atlantic Ocean around Cape Horn. Additional notable locations are the Eastern Coast of Brazil (700 miles long), SW coast of Chile (800 miles), Madagascar's Eastern Coast (800 miles), and the Wild Coast at Eastern Cape of South Africa (400 miles), the SW coast of Alaska, the SE coast of Canada, the NW coasts of Spain and Portugal, the Eastern coast of Australia, and the Gulf of Guinea in Africa.

Wave Characteristics and Power

An ocean wave constitutes the highest part called a crest and the lowest part above the unmoving water called a trough (Fig. 4.1). The vertical distance between the crest and the trough is the wave height and the amplitude is half of the wave height. A horizontal distance between two crests or two troughs is called a wavelength, while the time between the passing of two successive waves is the wave period. Similarly, wave frequency is the number of waves that pass through a certain point in a unit of time.

The power output from the Sun and the wind is measured per unit area, while wave power is measured per unit length and is a function of wave crest length (Fig. 4.2). For example, a wave can have an energy density of 50 kW/m in average North Atlantic Ocean conditions.

The power from waves per 1 m length of wave crest can be expressed using Eq. (4.1) (H is the wave height and T the wave period).

Fig. 4.1 Schematics of an
ocean wave

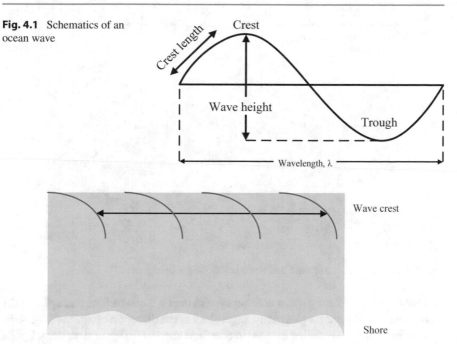

Fig. 4.2 Sketch of wave crest length along a shoreline

$$P - \frac{H^2 T}{2} \tag{4.1}$$

On a microscopic level, waves are composed of particles of water oscillating up and down. Near the surface, the paths of water molecules are the same size as the wave height, while the path length decreases below the surface. Approximately 95% of the wave energy is between the sea level to a depth equal to one-fourth of the wavelength $h = \frac{\lambda}{4}$. To capture the maximum energy from a wave, a device should come in contact with the majority of water particles in this layer (Fig. 4.3).

Wave Power Conversion

There are two types of wave conversion devices, onshore and offshore. There are also two principles of operation, one based on air turbine, whereby air is forced by the waves in and out of chamber containing a turbine, and the other based on a water turbine.

Oscillating Water Column (OTC) devices rely on wave power to push a column of air through a wind turbine. The rushing air causes the turbine to spin, creating electricity. As water fills the volume within the chamber, the air is forced upward out of the chamber and through a turbine. As water flows out of the chamber, the air is pulled down into the structure through the turbine.

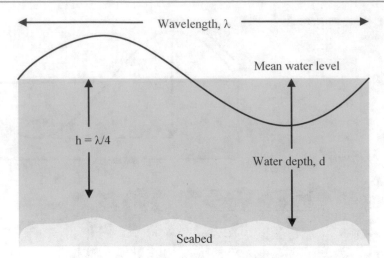

Fig. 4.3 Sketch of an ocean wave and water below the surface

One type of oscillating water column system uses a Wells turbine that spins in the same direction to produce electricity, regardless of the direction of airflow (Fig. 4.4).

One such turbine installed in 2000 in Scotland was the first grid-tied, commercial wave power system in the world. It comprises two 250 kW turbines and utilizes wave power in the range of 15–25 kW/m.

Another type of onshore device uses a hydroelectric turbine with a structure called tapered channel (TAPCHAN). The waves enter the structure on the land through a gradually narrowing or tapered channel with wall heights typically 3–5 m above the water level. The waves enter the wide end of the channel and propagate, which causes the rise of water at the narrow end, spilling of the wave crest over the walls, and filling of the reservoir (Fig. 4.5). The water is then released through a penstock to a conventional turbine, similar to the typical hydrothermal systems.

The onshore wave generation is a promising new technology with no associated emissions. However, the technology has an ecological impact by occupying a considerable length of coastline, proportional to the desired power output and inversely proportional to the average local wave energy density. The offshore wave power technologies, on the other hand, move the power production away from the shore, where the generating equipment has a far less significant impact on the local ecology. In addition, the wave power is the largest above continental shelf boundaries, where there is a sudden change in water depth.

There are a number of promising projects for floating devices used offshore. These energy converters can be classified, in terms of their location, as fixed to the seabed in shallow water, floating offshore in deep water, or attached to the seabed at moderate depths.

One such device, called Pelamis, which means "sea snake," resembles a shape of a snake, and it is several hundred meters long and 3.5 m in diameter. The device, comprising a number of cylindrical segments (e.g., four) on hinges, is anchored to

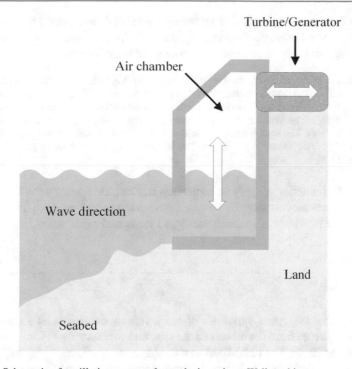

Fig. 4.4 Schematic of oscillating water column device using a Wells turbine

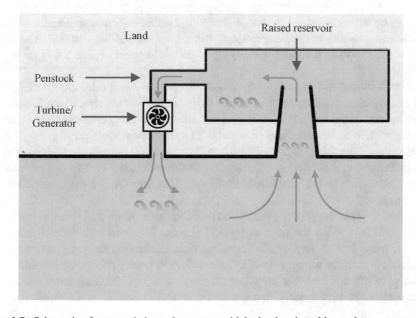

Fig. 4.5 Schematic of a tapered channel structure with hydroelectric turbine and a generator

the ocean floor and moves paralleling the movement of the waves. This also causes a movement of a hydraulic fluid in hinges which drives a generator and produces electrical power. Each hinge point can generate 250 kW of power.

> Another type of device (called Linear Generation Buoys) is located several miles offshore and about 30–70 m deep in water. The principle of operation of this device is based on magnetic linear generation. There are numerous versions of the design, but a typical device consists of a shaft anchored to the ocean floor and a free-floating cylinder moving up and down with waves. The shaft is a magnet, surrounded by an electrical coil, which moves like the waves lift and lowers the cylinder. The buoys are typically 4 m long, with 30 m between them.

Several conceptual devices rely on an underwater turbine whereby the floating portion of a device moves with the waves. Those movements produce a mechanical response in a piston or a specially designed hose and push water into a turbine (Fig. 4.6).

Tidal Power

The ocean tides are cyclic variations of sea levels on coastlines as a result of the gravitational forces from the moon and the sun, and the centrifugal forces due to the rotation of the Earth. The period between the high sea level, called "flood," and the low sea level, called "ebb," is 12 h and 25 min or half a lunar day. Other factors, such as ocean bottom topography, also contribute to the overall effect. Unlike waves, the tides are accurately predictable.

The tides on the global level are manifested as two bulges in the Earth's oceans, a larger on the side toward the moon due to gravitational attraction and a smaller on the far side due to the centripetal forces as a result of the rotation. The tides follow the moon as it rotates around the earth and appears twice a day on every coastline.

Superimposed to forces from the moon that causes lunar tides are gravitational forces from the sun that produce smaller tides because of the much larger distance from the sun to the Earth (Fig. 4.7). The combination of the earth–moon and earth–sun interactive forces produces seasonal tidal variations correlated with the phases of the moon. The tides are amplified when the moon and the sun are aligned during the full moon and new moon. During the first quarter, the two forces are out of phase, resulting in the neap tides and less difference at the opposite sides of the Earth.

When considering power generation from the tides, additional factors should be considered that affect the tides at specific shores. These include local winds and storms, the effect of the Coriolis forces on ocean currents, the temperature variations, depth, and the topography of the ocean floor. The variations in the ocean level because of the tides may be pretty small, roughly 0.5 m, far from the shore in the open waters but significantly higher near the shore depending on the topography of

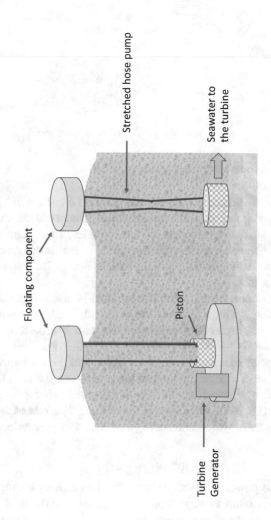

Fig. 4.6 Sketch of two types of wave power-generating devices anchored to the seabed and based on underwater turbines. A device with a piston on the left and with stretched hose pump on the right

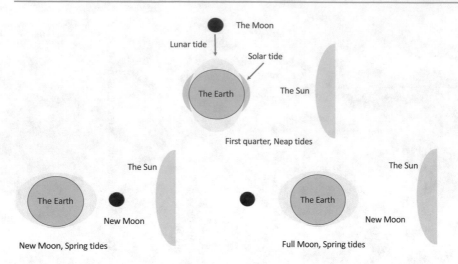

Fig. 4.7 The lunar and solar tides for different phases of the Moon

the ocean floor. In the case of shallow waters and closer to the shore, the tidal variations are greater and go up to 3 m. If the geographical features of the shore exhibit
tapered or gradually narrowing channels of the estuaries, for example, the tidal
amplitude can be much higher. For example, the tidal differences can reach 10–15 m
due to resonance effects in the narrow estuaries such as Severn Bore and Humber in
England and Hoogly in India.

The geographical features, as described above, can be used to construct reservoirs, called tidal barrages, which can store the rising tidal water and release it during the low tide through a hydroelectric turbine. As with hydroelectric dams, the
two most important factors for power generation are water flow and water head.
Barrages need to collect a significant volume of water, and the tidal variation must
be large enough to create a significant head of water between the high tide and the
low tide. The power, P, available from a body of water depends on the head in the
estuary, H_E, the acceleration due to gravity, g, the volume flow rate of the water, Q,
and an efficiency factor, η, which considers head loss and generation efficiency
(Eq. 4.2):

$$P = \eta \times 1000 \times Q \times g \times H_E \tag{4.2}$$

The energy the tidal power plant can produce from an estuary is proportional to
its area, A, and the maximum average depth of the estuary or the tidal range. The
mass of water depends on the estuary volume ($h \times A$) and density of water, ρ:

$$m = \rho \times h \times A \tag{4.3}$$

The potential energy stored in a tidal barrage is proportional to the stored mass
and average water height of the estuary (Eq. 4.4).

$$E = mgh = \left(\rho \times \frac{h}{2} \times A \right) \times g \times h = \frac{\rho \times h^2 \times A \times g}{2} \qquad (4.4)$$

The average power available from a tidal barrage station can now be obtained by dividing the expression for energy with the tidal period, T, (Eq. 4.5).

$$P = \frac{\rho \times h^2 \times A \times g}{2T}. \qquad (4.5)$$

The available energy is proportional to the square of the tidal range (i.e., tidal height difference). In practice, the average tidal range must be at least 5 m to generate sufficient power.

Tidal Power Generation

Tidal mills were used in Spain, France, and the United Kingdom close to a thousand years ago. The global potential for electricity production from tidal power is around 530 MW as of 2019. The two largest tidal power stations are La Rance in France and Sihwa Lake in South Korea, with almost identical net electrical energy output. La Rance, which was built in 1967, is 740 m long, and it has a maximum tidal range of 12 m and a typical water head of 5 m. It is obvious that the power is produced using low head turbines and a large volume of water. The power station has 24 reversible pump turbines with a net energy output of 480 GWh per year.

Besides the power generation using low head turbines during the low tide (i.e., ebb generation), the kinetic energy of a tidal current in an estuary can be used to power submerged turbines as the estuary fills with incoming water (also known as flood generation).

It is also possible to use the two methods combined and generate power during both ebb and flood cycles.

Environmental Impact of Tidal Generation

Tidal power generation using water-filling tidal barrages has a similar environmental impact as the hydroelectric dam power plants. The change in the natural ebb and flow of an estuary disturbs the sediment deposition and over time reduces its depth. The water also becomes more turbid and reduces the amount of light that penetrates to the bottom, which negatively affects the naturally rich ecosystem in the estuaries but can also have consequences on the local activities and economy.

The potential ecological problems with onshore tidal power generation are avoided by moving tidal power production offshore and using non-barrage technology. In deep waters, there is very little or no impact on the ecosystem because the underwater turbine blades rotate at a low velocity and are not affecting ocean species. Tidal generation is also more predictable than offshore wind power. The offshore turbines are anchored to the ocean floor and immersed in a tidal stream to produce power during the abb and flow of the tides.

Ocean Thermal Energy Conversion (OTEC)

The ocean thermal energy is based on the exploitation of the temperature difference gradient of 20 °C or more between the surface layers of water and deep water. The greatest temperature gradient is found in tropical and subtropical areas at a depth of 1 km or more. The process of converting the thermal gradient into electrical energy starts by using the warm water to vaporize a working fluid, which then expands and drives a turbine, before condensing back to liquid through heat exchange with cold water.

The three basic types of OTEC power plants include closed-cycle, open-cycle, and various combinations of the two. All three types can be built on land, mounted on offshore platforms anchored to the seafloor, or carried by ships that move from place to place. High capital investment and low conversion efficiency are the disadvantages of the OTEC technology that nonetheless has the resource potential to meet the global energy demand. OTEC has the potential to supply power much more efficiently than wave power.

In a closed-cycle process, a working fluid such as ammonia is first evaporated and then condensed back to liquid through a heat exchange with cold water. The working fluid remains in a closed system and is continuously recirculated. The requirement for the working fluid is to evaporate near atmospheric pressure at the available seawater temperatures. The overall principle of operation resembles that of refrigeration systems and is, therefore, well understood and scalable.

As of 2019, there are only two grid-connected OTEC facilities, one in Japan built in 2013 and the other one in Hawaii built in 2015.

Case Study: Wave Power in Portugal Portugal is showing remarkable advances toward developing renewable power generation. The country now derives over 70% of its electricity from renewable sources, primarily from hydropower, but its ocean power generation has remained relatively unexplored.

> If the current trend continues, Portugal can reach the goal of 100% renewable power, and ocean power has been expected to provide up to 25% of Portugal's total power consumption. Portugal's geography is ideal for harvesting ocean power based on the long coastline, and its power resource potential is estimated at 3–4 GW.

The ocean energy generation in Portugal has been under development for a long time, and the world's first wave farm was built there. In 2008, the wave farm started operation in Aguçadoura, located 5 km off the coast of north Portugal. The project was initiated by the Scottish company Pelamis Wave Power and had three Pelamis units installed, with a total capacity of 2.25 MW, sufficient to power more than 1500 Portuguese households at that time.

One of Portugal's newest ocean power generation projects is the 5 MW offshore wave energy generation project located near the port of Viana de Castelo. Future plans include the installation of 20 MW of wave power in four locations in Portugal.

Exercise 4

1. What are the advantages and disadvantages of harvesting ocean power?
2. What is a Wells Turbine? Write down its advantages and disadvantages.
3. Discuss the key differences between wave power and tidal power.
4. Make a comparative analysis between onshore and offshore ocean power. Include all types of ocean power available.
5. Draw schematic diagrams of open-cycle, closed-cycle, and hybrid OTEC systems.
6. In a certain tidal power station, the water density is 1030 kg/m^3. The usable tidal range is 8 m and the basin area behind the barrage is 20 km^2. What is the output power of the station? (145 MW).

Bioenergy

5

Introduction

Bioenergy is contained in biomass and its derivatives and converted to useful forms of energy. Biomass is usually described to be in the form of wood, crops, waste, and derivates such as gas and fuels (Fig. 5.1).

Biomass forms the surface layer of the Earth called the biosphere, and it is critical for maintaining the Earth's atmosphere and all life on Earth. The energy in the biosphere is constantly used and replenished from the energy of the sun.

Biomass from the biosphere can be burnt directly to produce heat, generate power, or processed to make liquid fuels. The fuels can be directly used for many applications or through further processing most of the fuels can also lead to pathways to generate electricity. A diagram of the generalized routes to process biomass is shown in Fig. 5.2.

Historically, wood has been burnt for heat and cooking for over 10,000 years, while plant and animal oils have been used for centuries to provide lighting (e.g., kerosene as a lantern fuel). The extent of the use of traditional biomass is difficult to measure, but a very rough estimate is that it contributes 10% to the total world energy supply.

The energy in the biomass, such as flowering plants or trees, is chemical energy converted from solar energy through the process of photosynthesis. The processes for conversion of the chemical energy to useful energy include direct combustion to produce heat, thermochemical conversion to produce various fuels, chemical conversion to produce liquid fuels, and biological conversion to produce liquid and gaseous fuels.

© The Author(s), under exclusive license to Springer Nature
Switzerland AG 2021
E. Hossain, S. Petrovic, *Renewable Energy Crash Course*,
https://doi.org/10.1007/978-3-030-70049-2_5

Fig. 5.1 Diagram of the
main types of biomass

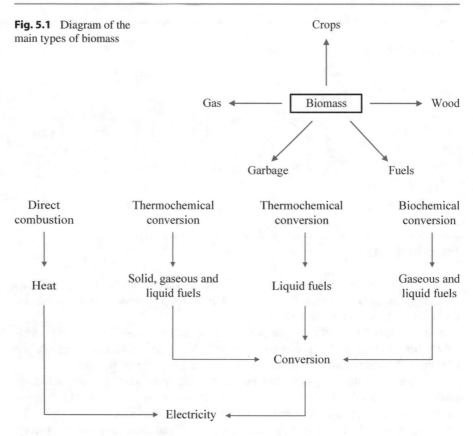

Fig. 5.2 A diagram of routes for biomass processing

The Origins of Bioenergy

Biomass is generated from a relatively small portion of the overall solar radiation on the Earth. The total yearly solar radiation on the Earth is about 5.4×10^6 EJ and from that only 0.023% or 1260 EJ is used by the plants in the photosynthesis reaction. The process starts with light absorption by the substance chlorophyll present in the leaves of the plants and the use of absorbed energy to enable reaction with carbon dioxide and water (Fig. 5.3).

The products of this reaction are sugar glucose (carbohydrate) and oxygen (Eq. 5.1). The reaction is endothermic which means that it requires the input of energy from the environment.

$$6CO_{2(g)} + 6H_2O_{(l)} + Sunlight \rightarrow C_6H_{12}O_{6(aq)} + 6O_{2(g)} \tag{5.1}$$

In this reaction, the energy from the sun is converted to chemical energy in the carbohydrate used as a fuel in biological systems to provide growth of biomass. The basic principles of chemical reactions and thermodynamics indicate that this chemical energy contained in plants can be released back to the environment in a reverse

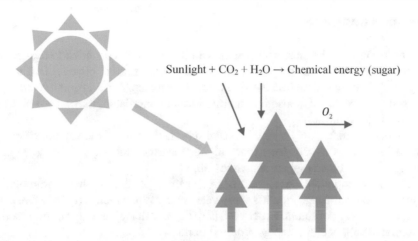

Sunlight + CO_2 + H_2O → Chemical energy (sugar)

O_2

Fig. 5.3 Schematic of the sunlight conversion in the plants through the photosynthesis reaction

exothermic reaction. If such a reaction is controlled and the energy is captured, it can be used as thermal energy for heating or further converted to electrical energy. As a result of the overall process, the photons of light are converted, through a series of steps, sometimes taking a long time, to electricity.

The photosynthesis reaction, endlessly replenishing the biomass, is responsible for the growth and maintenance of the surface layer of the Earth called the biosphere. Biomass decays naturally over long periods of time. Under the pressure and temperature conditions, together with dead animal remains, it converts to a form known as fossil fuels, i.e., coal, crude oil, and natural gas. The chemical energy contained in the biomass can, therefore, be recovered and used from the fossil fuel deposits millions of years old, or it can be used as well by directly removing it from the biosphere to produce heat in a reaction with oxygen or further converting it to electricity.

Biomass comprises three components, cellulose (40–60%), hemicelluloses (20–40%), and lignin (10–25%). Lignin is a polymer compound that provides structural integrity to plants and, in combination with hemicellulose, provides a protective layer around the cellulose. Lignin is left as a residual after the sugars in biomass are fermented to ethanol, and it is sometimes described as a sulfur-free coal. Another plant constituent, used for energy storage, is starch.

All these belong to the group of organic compounds called carbohydrates, composed of hydrogen, carbon, and oxygen and used in plants for basic functions, such as energy storage and transport, and ultimately the growth. Simple sugar is a monosaccharide, while larger carbohydrates composed of many monosaccharides are polysaccharides that include starch, cellulose, and hemicellulose. Due to weaker intermolecular bonding, starches are more easily broken down into glucose making them suitable to be used for energy storage in plants. Cellulose, on the other hand, is rigid and provides structure to the plant.

Biomass Resource

Biomass contains chemical energy that can be potentially converted to heat energy or electricity. While clearly not an unlimited energy source, biomass is a perfect renewable energy source because of the constant replenishment from the energy of the sun and because it is also not intermittent and unreliable, like solar, wind, and ocean energy sources.

Despite capturing only a small portion of solar radiation, the amount of energy stored in biomass is still many times higher than the global energy consumption. The facts about bioenergy are shown in Table 5.1.

The most important conclusion from the table of facts is that the amount of energy stored in the land biomass every year is 95 TWh, which is roughly five times the global energy consumption per year, and it means that bioenergy has the resource potential to supply entire energy required in the world.

Using biomass or conversion of bioenergy by simply burning it for heating has been the most common energy consumption method for thousands of years, and it is estimated to still be one of the largest contributors to total global renewable energy used today. The fact that a low-tech energy conversion method surpasses advanced technologies in global contribution simply means that biomass is widely used for heating, lighting, and cooking in developing countries. It is important to note that the information about using biomass for heating and cooking is not as reliable as for the other technologies because of its almost completely distributed nature and relying on estimates.

The use of biomass to generate electricity or produce fuels for transportation and other purposes is inevitably compared to the use of fossil fuels. The energy content comparison between biomass, chemically carbohydrates, and fossil fuels, which are hydrocarbons, composed of hydrogen and carbon, is shown in Table 5.2. The data, which represents typical energy content, clearly expose that fossil fuels are a more concentrated source of energy than biomass, a result of a biomass compression process over millions of years.

Beyond the unsystematic utilization of naturally occurring biomass (trees, branches, etc.), and some semi-organized collection of lumber residues, the biomass can be utilized for the production of electricity and fuels. Such biomass includes

Table 5.1 Bioenergy facts

Bioenergy related facts	Value
Total mass of the living matter	2000 billion tonnes
Total mass of land plants	1800 billion tonnes
Total biomass of forests	1600 billion tonnes
Energy stored in terrestrial biomass	25,000 EJ
Net annual production of terrestrial biomass	4×10^{11} tonnes/year
Annual rate of energy storage by land biomass	3000 EJ/year (95 TW)

Table 5.2 Energy content for selected fossil fuels and biomass

Fuel source	Typical energy content, GJ/tonne
Natural gas (methane)	55
Petroleum (crude oil)	48
Coal	28
Charcoal (residue from heating wood)	30
Dry wood	18
Straw (byproduct after grain removal)	15
Green wood	6

forest debris, scrap lumber, some manure, and waste. The most important, however, is the use of certain engineered crops. If such a crop is produced on 1 ha of land (10,000 m²), it receives roughly 1500 kWh/m² or 5.4 GJ/m²/year for a location with a moderate average solar irradiation of about 4 kWh/m²/day. However, it can be shown that there are several energy absorption and conversion steps, each with limited efficiency. The final stored chemical energy in the crop is only 0.64% of the sun's energy reaching the site or about 350 GJ for the whole 1-ha location.

Despite low overall conversion efficiency, the potential for biomass utilization for electricity generation is significant. Biomass generation is roughly carbon-neutral because CO_2 would be emitted into the atmosphere also through natural plant decay. It is also sustainable if the generation of biomass energy by utilizing the energy of the sun in the photosynthesis reaction is equal to or larger than the consumption of the biomass in the electricity generation. While fossil fuels are a form of biomass that lived millions of years ago and a long time was required for the conversion, the use of fossil fuels to produce energy is clearly not renewable because the previously formed reserves are used that cannot be replenished. Another form of biomass that is not considered a renewable energy source is peat, an incompletely decayed vegetation formed as part of the soil over hundreds of years.

Despite some arguments that biomass is not a completely clean energy source and that it emits large amounts of CO_2 due to low conversion efficiency, it should be understood that for the Earth's atmosphere balance biomass is carbon-neutral because the same amount of CO_2 is used in the photosynthesis reaction (Fig. 5.4). In addition, biomass can be replenished in a short time, so it will never be depleted as fossil fuels will. As a result, biomass can replace the use of fossil fuels not only for power generation but also replace transportation and other fuels with biodiesel and ethanol, while methane (natural gas) can be collected from municipal waste.

Biomass for Energy

Biomass is converted to energy, heat, electricity, and fuel through two main methods, thermochemical and biochemical. Both methods can lead to the generation of all three types of energy. The simplest method is combustion or burning of biomass to produce heat. Combustion can also be used to first heat water to steam, which is then used in a steam turbine to generate electricity (Fig. 5.5).

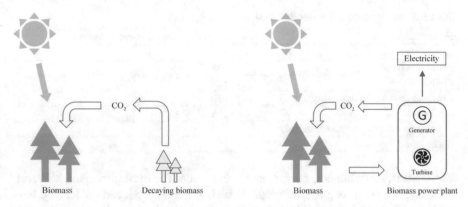

Fig. 5.4 Schematic of the natural carbon cycle (left) and carbon dioxide balance in case of biomass use for electricity generation (right)

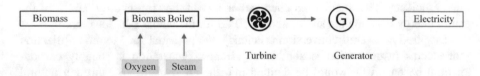

Fig. 5.5 5 Schematic of a biomass combustion process to generate electricity

The oxidation reaction of carbohydrates results in the formation of CO_2, H_2O, and the release of heat, which is the difference between the chemical energy of the products and the reactants. One type of carbohydrate is the simple sugar glucose, the constituent of starch and cellulose. The chemical reaction in which glucose is oxidized produces heat but also CO_2 and water (Eq. 5.2).

$$C_6H_{12}O_6 + 6O_2 \rightarrow 6H_2O + 6CO_2 + Heat \qquad (5.2)$$

Note that glucose oxidation is a reverse process of photosynthesis (Eq. 5.1).

Gasification is a process in which biomass, or other solid materials such as coal, is reacted with steam and oxygen at very high temperatures (>1000 °C) and pressures of up to 30 atm. The product is a gas composed of carbon monoxide, hydrogen, and some methane. The gas, with an energy content of 3–5 MJ/m³ (about ten times lower than natural gas), can then be used to produce heat from a gas turbine to provide fuel for fuel cells, or after some conversion it can be used as a gaseous fuel for an engine.

Biomass can also be liquified through a process of pyrolysis consisting of heating while carefully controlling the air supply to minimize gasification and only increase biomass volatility. The volatile components, which contain more energy than solid residue, are collected and then condensed to produce liquid fuel, i.e., biooil, with about half of the energy of crude oil. The smaller quantity of gas produced can be used in the same way as described in the gasification process to power a gas turbine or a fuel cell. The solid residue component is called charcoal, and it is used as fuel as well. Liquid fuels can be produced from biomass also through a process called solvolysis that involves the use of organic solvents at 200–300 °C.

Biomass can be converted using biochemical processes, digestion, and fermentation, a chemical breakdown of a substance using enzymes and bacteria to produce ethanol, and use it as a fuel. Anaerobic digestion is carried out in the absence of oxygen and in the presence of microorganisms to generate gas, which can then be used, after additional synthesis, to produce electricity in a gas turbine or a fuel cell.

In several biomass conversion routes, an intermediate product is a gas, composed mainly of carbon monoxide and hydrogen and called synthesis gas. Further processing of the synthesis gas at high temperature and pressure over a catalyst (Fischer–Tropsch process) leads to a mixture of liquid and gaseous hydrocarbons (Fig. 5.6). The gases are used for heating, and the liquids can be refined into fuels used for vehicles.

Energy Crops

Energy crops are purposely grown for use as fuel or conversion into other biofuels. Because of the poor efficiency of converting sunlight into stored biomass energy, farming of energy crops requires the consumption of large land areas. There are two types of energy crops, forestry crops and agricultural crops.

Forestry crops are fast-growing and dense trees harvested to be used as fuel and burnt to produce steam for heating or to drive generators. Harvesting of forestry crops can be done every few years and typically has a yield of 10 tonnes/year.

Agricultural crops are grown for conversion to liquid fuels such as ethanol. The most important agricultural energy crops and their yields are shown in Table 5.3.

The energy crops that contain sugar are suitable for direct fermentation and conversion in biochemical route to ethanol, which has a high energy content of 0.024 GJ/L and is used as a liquid fuel. Some of the energy crops used to produce ethanol include sugar cane, maize, cassava, and wood. The yield of 10% ethanol depends on the location, and it varies from 160 to 12,000 L/ha/year. Agricultural crops such as canola, mustard, sunflower, soybeans, and corn may be processed for their oil and converted to biodiesel.

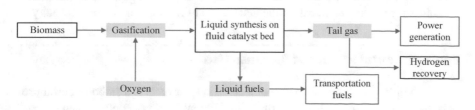

Fig. 5.6 Diagram of the Fischer-Tropsch process for the production of transportation fuels from biomass feedstock. The process is also used for coal

Table 5.3 Energy crop yields

Energy crop	Yield, tonnes/ha/day
Sugar cane	35
Maize	15
Bagasse	10
Wood (temperate region)	10
Wood (tropics region)	20

Waste Biomass Sources

Waste products such as crop wastes, animal wastes, and municipal solid wastes (MSW) have a great potential for conversion to heat, electricity, or fuel. For example, temperate (grown in mild temperatures) crop wastes from corn, barley, and wheat have a global potential of up to 20 EJ annually, which currently has been barely exploited. A significant portion of the tropical crop wastes from sugar cane, rice, or bagasse has been used in gasification processes to produce heat or electricity.

The municipal solid waste has the potential to generate 9 GJ/tonne through anaerobic digestion to produce biogas or fermentation to produce ethanol, while methane from animal wastes can be captured to generate power.

Biomass Plant

A biomass plant is constructed to handle different types of biomass such as wood, animal manure, municipal waste, and energy crops (forestry and agricultural). A single plant may not be able to accept all types of biomass.

The first step in the treatment process is drying, which can be done in the ambient air, using forced air drying, kiln drying (controlled air circulation and humidity), or torrefaction (drying in an inert atmosphere at 200–300 °C).

The next step is to prepare and form the biomass by resizing it into a uniform shape. This is done by combining several mechanical steps, sawing, splitting, chipping, shredding, and grinding. It is then followed by densifying treatment such as pelletizing, briquetting, and cubing.

The plant design from here differs based on the type of the process used, i.e., direct combustion, thermochemical conversion, chemical or biological conversion. Based on the process, the final products include combustible gas, liquid fuels, ethanol, biocrude oil, and charcoal, or the generation of thermal energy or electricity.

Environmental Impact of Biomass Use

Although the use of biomass to generate electricity or fuels is a carbon-neutral process, the impact of bioenergy on carbon dioxide emission is a controversial subject. The exploitation of biomass generates much less CO_2 than burning fossil fuels, but the deforestation that occurs when growing biomass reduces the number of plants for CO_2 sequestration. Even if the biomass crops are harvested on previously deforested land, they usually require irrigation using energy, which offsets the amount of electricity generated.

Bioenergy power plants have large footprints, and for 10 million kWh between 300 and 1000 ha are required, compared to 40 for solar PV generation and 100 ha for a wind turbine farm to produce the same amount of electricity.

Case Study: Miscanthus Miscanthus is a non-woody energy crop, with a thick woody stem and a low water content. Miscanthus yields up to 83 tonnes/ha (30 dry tons/acre)/year. Because of the low water content, it is suitable for direct combustion to produce heat but also for ethanol generation.

Assuming that ethanol yield of 100 gallons per dry ton (420 L/tonne) is possible, it would lead to 3000 gal/acre or 28,400 L/ha. In the United States, there are 450 million arable acres of land (about 18% of the total land area), and using, for example, 100 million acres would produce about 300 billion gallons (1136×10^9 L)/year of ethanol. The total US consumption of gasoline per year (2019) was 142 billion gallons, which can be replaced with roughly 200 billion gallons of ethanol. In addition, diesel consumption is 47 billion gallons (2019).

In summary, using the arable area of 100 million acres or roughly 4% of the US land area to grow miscanthus and produce ethanol could provide more than the total US demand for transportation fuels, gasoline and diesel.

Exercise 5

1. What are the advantages and disadvantages of using biomass energy?
2. Discuss the difference between biomass, biofuels, and biogas.
3. In a coal-fired powerplant, coal is burnt to produce heat. Instead of coal, if biomass is burnt, what would be the benefits and drawbacks of the replacement? Check out the story of the Ironbridge Power Station in the UK in this regard.
4. How is biomass different from fossil fuels?
5. How can biomass be established as a reliable and safe source of energy?

Geothermal Energy

6

Introduction

Geothermal resources are expected to make a significant contribution to the overall renewable energy mix. The energy can be used for electricity generation and space or water heating in single homes or district heating for residential or commercial applications.

Geothermal resources such as hot springs were used for heating in North America more than 10,000 years ago. The springs were used for heating and bathing, while minerals in the hot water brought healing effects.

Modern use of geothermal resources involves electricity generation by extracting heat from the depths of several kilometers.

Origins of Geothermal Energy

When the Earth formed roughly five billion years ago, its interior was heated from the energy produced in the impact of astronomical bodies. This energy is not being replenished, and it is slowly diminishing. Since at present the Earth is warmer than expected, it is clear that there is another source of heat. This additional heat, about 50% of the total heat, is generated in the decay of radioactive isotopes with long half-lives, Thorium 232, Uranium 238, and Potassium 40.

The heat generated in the inner layers (Fig. 6.1) flows to the surface because of the temperature difference of over 4000 °C between the surface and the inner core (Fig. 6.2).

Heat transfer from the interior of the Earth to the surface occurs through convection in the liquid rocks of the inner core, outer core, and mantle, and closer to surface, the heat is transported through conduction, a process where atoms and molecules collide. In the Earth's crust, which is the solid rock, the heat is transferred through vibrations.

© The Author(s), under exclusive license to Springer Nature
Switzerland AG 2021
E. Hossain, S. Petrovic, *Renewable Energy Crash Course*,
https://doi.org/10.1007/978-3-030-70049-2_6

Fig. 6.1 Cutaway profile
of the Earth layers

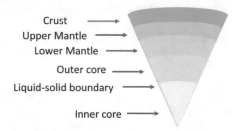

Fig. 6.2 The Earth
temperature as a function
of depth

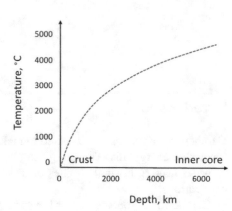

Geothermal Resource

The steady amount of heat that reaches the Earth's surface is 10^{21} J/year, which is
about one thousand times lower than the energy from the sun but still about 2.5
times the world energy demand. And, while the global geothermal energy content
within the earth is tremendous, it is not infinite for a specific location and it can be
exhausted, never to be renewed on the scale of a human lifetime.

The steady geothermal heat flow from the surface is insufficient for generating
power because the average density is 60 mW/m^2, compared to the solar energy density of 1000 W/m^2. However, the diffusion of geothermal energy is not uniform
everywhere on the Earth, and there are geothermal resources of low, medium, and
high enthalpy (or heat). A maximum heat flow as high as 300 mW/m^2 can be
obtained in some areas affected by the anomalies in the Earth's crust due to tectonic
plate boundaries and geological characteristics. Active plate boundaries due to volcanic activities typically have high heat flow and also offer natural features such as
boiling pools, geysers, and volcanic vents.

The tectonic (or lithospheric) plate movement is critical for creating fractures in
the Earth's crust through which higher amounts of heat could come to the surface.
The Earth's crust is divided into 13 major lithospheric plates, and regions of high
geothermal heat resource potential are located along the plate boundaries, which are
connected to form the so-called Ring of Fire. It is a path along the Pacific Ocean
known for common earthquakes and active volcanos (Fig. 6.3).

Fig. 6.3 Sketch of the lithospheric plate boundaries (at night) forming the Ring of Fire path

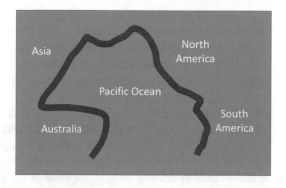

In North America, for example, a small part of the Ring of Fire extends from British Columbia to Northern California and is known as Cascadia Subduction Zone, a long corridor prone to frequent earthquakes and probable volcanic eruptions. In the areas like this on the Earth where lithospheric plates move against each other, it is possible for the hot liquid magma to ascend through the fracture and to carry high amounts of heat relatively close to the Earth's surface. In some especially active areas such as the Yellowstone National Park in the United States, the geothermal heat freely reaches the surface in the form of geysers, i.e., hot springs where water and steam lift into the air at high pressure. There are many other similar natural wonders along the Ring of Fire in Japan, the Philippines, Central America, the Great Rift Valley of Africa, and Iceland.

Geothermal Sites

Geothermal sites, i.e., heat resources, are classified into three categories based on their enthalpy (the total heat content of a system):

1. Low enthalpy with temperature less than 100 °C.
2. Medium enthalpy with temperatures ranging from 100 °C to 180 °C.
3. High enthalpy with temperatures higher than 180 °C.

The amount of heat transferred from the hotter inner layers of the Earth to the surface depends on the temperature difference; the thermal conductivity of the rock, k; depth from which the heat is coming, d; and surface area under observation, A, (Eq. 6.1).

$$Q = \frac{kA\left(T_{hot} - T_{cold}\right)}{d} \tag{6.1}$$

In a classical example that represents a typical site, the temperature is 63 °C at a depth of 2 km, the surface temperature is 15 °C, and the thermal conductivity of the rock is 2.5 W/m °C, giving the heat flow of 0.060 W/m².

High enthalpy resources are mainly used for generating electricity, while medium and low enthalpy sites are typically used for direct application of heat from heating single houses or district heating facilities, to greenhouses, aquaculture, snow melting, and pools. In some locations, the whole cities (e.g., Reykjavik in Iceland) are heated from the geothermal plants processing hot water from geothermal wells and transferring it through pipes to individual users. The city of Klamath Falls in Oregon, USA, located along the Cascadia Subduction Zone corridor, and its well-known university, Oregon Institute of Technology, are largely heated by geothermal water.

The geothermal heat resource extraction is not, however, limited to capturing heat that naturally penetrates the crust and comes to the surface. It is also possible to extract heat by drilling through the crust to a depth of several kilometers to reach reservoirs of molten rock and extract high heat or water vapor. This type of heat extraction is used for generating electricity in an efficient process because of the large temperature difference that can easily be 1000 °C, providing high-value heat.

The sites suitable for geothermal heat extraction comprise aquifer, caprock, and heat source. The aquifer is the layer that contains water and is between the impermeable crystalline rock and impermeable cap rock (Fig. 6.4).

The porosity is an important requirement for the aquifer rock, usually a limestone, because it must allow permeation of water from the surface and to contain water in interconnected pores of the reservoir. The aquifer can be confined when it is beneath the caprock, and the impermeable rocks prevent loss of water, or unconfined, when it is open to the surface, and the rainfall infiltrates to replenish the aquifer.

The heat flows from the heat source below, and it is transferred through an impermeable rock through conduction to heat water in the aquifer. The cap layer above aquifer rock is also impermeable, trapping the heat in the aquifer. When water becomes heated, it may escape as steam or hot water through natural openings in the caprock called fumaroles, geysers, and hot springs or through man-made wells. Geothermal resources include both dry heat and hot water, which is more suitable for transporting it to the surface.

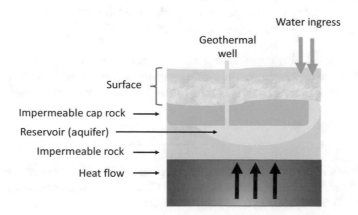

Fig. 6.4 Schematic depiction of the sub-surface cross-section of the Earth and unconfined aquifer

Geothermal Power Plant Technologies

Besides the heat that naturally ascends to the surface, geothermal energy, in the form of steam or hot water, can also be accessed by drilling the Earth's crust, from shallow depth to several kilometers. The geothermal heat brought to the surface is used to generate electricity in conventional steam turbines, where it is converted to mechanical energy used to spin the turbine and generate electricity. After exiting the turbine, the steam is exhausted to the atmosphere.

The majority of geothermal resources have only dry heat that can be used. The heat is exploited from these sites by injecting water deep below the surface into the high-temperature rock where it is heated to steam and brought back to the surface to power a turbine. The condensed water that exits the turbine is injected back into the rock. The water is heated again and available for the next cycle (Fig. 6.5).

Flash steam power plants are used when the geothermal resource is hot water (>180 °C) under pressure. Fluid is then pumped to the surface and into a tank at lower pressure, causing the fluid to rapidly vaporize (or flash) and turn a turbine, which drives a generator (Fig. 6.6).

Most geothermal power plants are Binary Cycle type used when the geothermal resource is hot water at moderate temperatures (<180 °C). As the hot water enters the power plant, it passes through a heat exchanger together with a secondary fluid with a lower boiling point, causing it to rapidly vaporize and drive the turbine. In this process, the corrosive geothermal source water never comes in contact with the turbine components.

Hot Rock Energy power plants are used when the geothermal resources do not have steam or water. The heat from dry rocks is extracted by pumping cold water under high pressure deep into the rock and then returning hot water to the geothermal plant.

Fig. 6.5 Schematic of a dry steam power plant

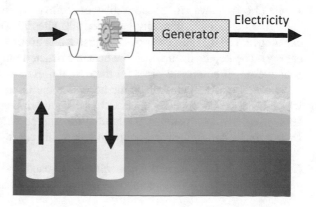

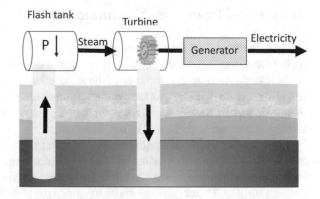

Fig. 6.6 Schematics of a flash steam power plant

Environmental Aspects of Geothermal Energy

The biggest environmental concern when using geothermal energy to generate electricity is bringing pollutant gases such as CO_2, H_2S, SO_2, and CH_4 to the surface. One of the most serious concerns is the highly toxic hydrogen sulfide gas, and sometimes, power plants have H_2S removal capabilities.

Hot water used from geothermal resources sometimes contains large amounts of silica, chlorides, and carbonates along with traces of heavy metals and when used in geothermal power plants can cause corrosion.

In some locations, long-term effects may cause gaseous pollution and induced seismicity.

Case Study 1: The City of Klamath Falls and Oregon Institute of Technology The city of Klamath Falls, in Southern Oregon, USA, located along the Cascadia Subduction Zone corridor, is a Known Geothermal Resource Area (KGRA). The city uses geothermal water to provide district heating services, residential and local greenhouse applications, and heating for sidewalks and bridges.

The geothermal water used has a temperature from 200 °C to 220 °C, considered a high-enthalpy resource site. Hot water is transported from the production well 1 mile (1.6 km) away to a heat exchanger facility, where clean water is re-injected back into the aquifer. The city has over 500 sites like this, called downhole heat exchangers (DHE), and each well typically has around 0.8 MW$_t$ (thermal or heat energy). The depths from which the heat is extracted range from 150 to 450 m. The heat exchanger comprises pipes for the reinjection of clean water.

The City of Klamath Falls is the home for the Oregon Institute of Technology, which in 1975 opened a Geo-Heat Center for studying geothermal energy. The university has a unique campus geothermal heating/cooling system and is in general known for its pre-eminent programs in renewable energy.

Case Study 2: Geothermal Aquaculture Aquaculture is a unique application of geothermal energy, where geothermal heat is used to heat water suitable for aquatic plants and animals and help increase the growth and production of aquatic species and destroy harmful organisms to keep the water body safe for desired species. Geothermal water is typically utilized in ponds, tanks, and raceways.

A Geothermal Aquaculture Research Foundation (GARF) located in Boise, Idaho, USA, undertakes developments to advance methods to utilize geothermal water to keep waters warm in coral farming, decorative water gardening, farming fishes and shellfish, keeping waters clean from unwanted aquatic weeds and organisms, and algae control. It has been demonstrated that warm water can improve the growth of fishes and ideal breeding conditions.

Exercise 6

1. Explain the origins of geothermal energy and exploitation technologies.
2. A commercial geothermal field of 4 km² is being developed in the Imperial Valley, CA. What is the total heat flow for this development in MW_h if the heat conductivity of the rock is 4 W/m °C, the surface temperature is 22 °C, and the heat source is at the depth of 2 km where the temperature is 92 °C? [Answer: 0.8 MW].
3. Discuss the advantages and disadvantages of geothermal energy.
4. How can geothermal aquaculture improve fish farming?

Solar Thermal Energy

7

Introduction

Solar thermal energy has been used earlier in history than any other renewable energy technology, and evidence exists for focusing mirrors being used in Mesopotamia to light fires several centuries BC. It is perhaps not surprising that the civilization that invented writing, mathematics, batteries, metallurgy, and hydraulic engineering also explored the use of solar power. Setting aside unproven legends such as the Archimedes' burning mirrors, it is clear that solar power has fascinated people from early ages and that it was directed toward demonstrating the thermal energy from the sun.

Several thousand years later, in the sixteenth and seventeenth centuries, focusing lenses and mirrors were used again to concentrate solar power. Leonardo Da Vinci proposed the construction of a concave mirror of large size on the side of a hill to be used to heat objects in the focal point.

At the beginning of the twentieth century, large-scale solar power was used for irrigation in Egypt. Solar-powered pumps were also used in Arizona and thermosyphon, low-cost water heaters were invented in California. In the 1930s, before the introduction of low-cost fossil fuels, about 80% of all homes in Florida had solar water heaters. It is estimated that presently over 60 million square meters of solar collectors exit in the world.

The Origins of Solar Energy

Solar radiation that reaches the Earth is generated in a nuclear fusion reaction in the sun. The process involves two hydrogen isotopes, deuterium with one neutron in the nucleus, and tritium, with two, that react to produce a helium atom and a neutron. The reaction releases large amounts of nuclear energy as heat (Eq. 7.1).

E. Hossain, S. Petrovic, *Renewable Energy Crash Course*, https://doi.org/10.1007/978-3-030-70049-2_7

$$\,_1^2H + \,_1^3H \rightarrow \,_2^4He + \,_0^1n \tag{7.1}$$

The energy from the nuclear reaction is first conducted from the core of the sun, at approximately 15 million degrees Celsius, to the surface, at temperatures of about 6000 °C. From the surface of the sun, the energy is radiated in all directions, and only a small portion reaches the Earth's atmosphere. The radiation then passes through the Earth's atmosphere, where it is attenuated and at the surface of the Earth is, on average, 1 kW/m². The radiation is composed of the spectrum of wavelengths based on the temperature of the surface of the Sun (Fig. 7.1).

Most of this radiation (about 48%) is from the visible spectrum with a wavelength between 0.38 and 0.78 μm. Around 7% of the sun's emission is in the range of 0.1–0.4 mm (UV range), and about 43% of the solar radiation is between 0.71 and 4.0 mm (near-infrared 0.71–1.5 mm, far infrared: 1.5–4.0 mm).

The upper curve in Fig. 7.1 refers to AM0 or extraterrestrial radiation. AM0 stands for "air mass of zero" and defines the solar radiation intensity at the edge of the Earth's atmosphere. This solar power density for AM0 is 1367 W/m² and is called the solar constant. The atmosphere of the Earth blocks a portion of the solar radiation (lower curve). The curve is not smooth because the radiation of different wavelengths is absorbed in the gasses and water vapor in the Earth's atmosphere to a different degree.

Solar Thermal Systems

Solar thermal systems that utilize heat from solar radiation can be passive or active. Passive systems include greenhouses, Trombe wall, and daylighting. The active systems can be concentrating or non-concentrating (Fig. 7.2). There are several other ways to classify solar thermal systems.

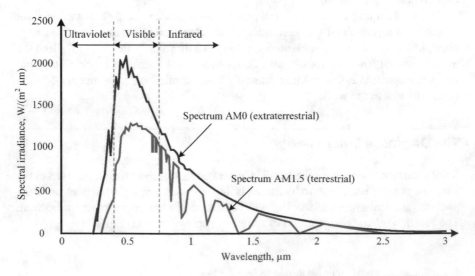

Fig. 7.1 Conceptual depiction of the spectrum of solar irradiance, i.e., the power density of solar radiation as a function of wavelength. The actual values and shape of the curves may be different

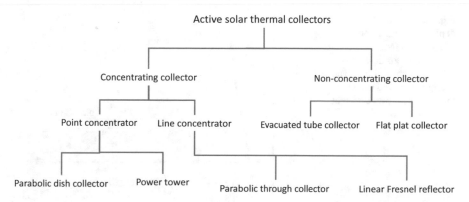

Fig. 7.2 Classification of active solar thermal systems

The passive solar heating applied to a building is based on a simple principle of absorbing maximum heat from the solar energy and preventing the structure from radiating heat away. The three main practices to achieve this goal are described here.

Greenhouses are designed to collect and preserve the heat. The principle of operation can be understood by combining the knowledge of the spectrum of solar radiation (Fig. 7.1) and the connection between the temperature of an object and the peak wavelength of its radiation, expressed in Wien's displacement law (Fig. 7.3).

The graph illustrates that objects at different temperatures emit spectra at different wavelengths, and the peak radiation is inversely proportional to the object's temperature, i.e., objects at higher temperatures radiate more energy and at shorter wavelengths. Wien's Law gives the wavelength of the radiation as a function of the absolute temperature (Eq. 7.2).

$$\lambda_{max} = \frac{b}{T} \tag{7.2}$$

In this equation, b is the Wien's displacement constant and T is the absolute temperature. If the wavelength is expressed in µm, the constant b is 2898 µm × K.

The solar radiation is emitted from the surface of the Sun at 6000 °C and has the maximum intensity wavelength in the visible region (approximately 483 nm), while a house, for example, is at 300 K and the maximum intensity wavelength from that object is in the infrared range (approximately 966 nm), far right on the graph in Fig. 7.3. Finally, to put it all together, the transparent roof and walls of a greenhouse are made to transmit short-wavelength radiation and prevent the passage of the long-wavelength radiation, effectively restricting the heat loss and heating the interior (Fig. 7.4).

A Trombe or a solar wall is a passive solar thermal method comprising an outside glass, an air gap, and a wall with a high thermal mass. When the sunlight is available during the day, the radiation passes through the glass, and thermal energy is absorbed by the wall (Fig. 7.5). The wall is painted black to absorb more sunlight, and it has a high heat capacity. The absorbed heat is stored in the wall and gradually radiated into the interior of the structure at night.

Fig. 7.3 Conceptual depiction of the intensity of light as a function of wavelength for objects at different temperatures. The actual values and shape of the curves may be different

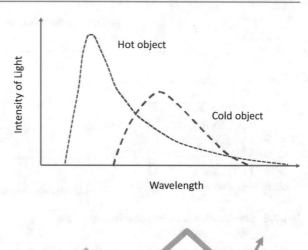

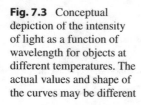

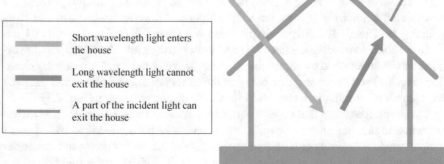

Fig. 7.4 Sketch of a greenhouse concept

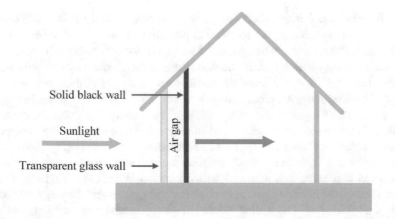

Fig. 7.5 A sketch of a Trombe (i.e., solar) wall

Finally, the daylighting method means optimization of the window glass for maximum light absorption and minimum radiation out of the structure to ensure heat retention within the structure. The requirements for an efficient glass include transparency for the visible radiation, minimum reflection loss, and translucency for long-wavelength infrared radiation (8–24 μm). Most of the efficient windows are double pane, with air in between.

Active solar thermal energy systems use solar thermal energy to heat a working fluid, usually water, that can be then used to store heat in a tank and further distribute heat. The fluid can be simply transported through the pipes or actively pumped. Non-concentrating active solar thermal systems use sunlight for direct heating of the working fluid while concentrating systems use mirrors and lenses to concentrate sunlight on a small area.

Non-concentrating collectors produce heat at moderate and low temperatures. Two types of non-concentrating solar collectors are flat plate collector (FPC) and evacuated tube collector. A flat-plate collector is used mainly for building heating and industrial process heat (Fig. 7.6).

The panel construction uses a similar principle as described above to capture most of the heat. It has a black surface, and the glass properties ensure that radiative losses are minimized. There are different patterns of tubes that carry water or working fluid. These types of collectors have about 50% overall efficiency in converting solar radiation to heated fluid.

Evacuated tube collectors have tubes made of borosilicate glass, 75% SiO_2 and 8–12% B_2O_3, or soda-lime glass, 70% silicon dioxide, 15% soda (sodium oxide), and 9% lime (calcium oxide). The tubes range from 2.5 to 7.5 cm in diameter and from 15 to 25 m in length. The evacuated tubes house absorber plates and copper pipes filled with internal fluid (Fig. 7.7). The heat is transferred from the absorber to the copper tube without losses in a vacuum, and the internal fluid in the copper pipes

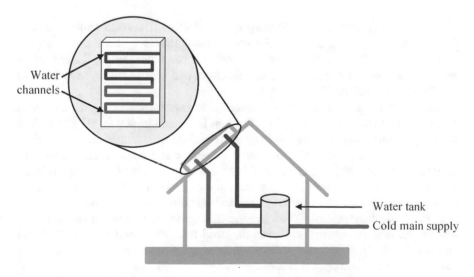

Fig. 7.6 Conceptual sketch of flat-plate solar thermal collector used to heat water

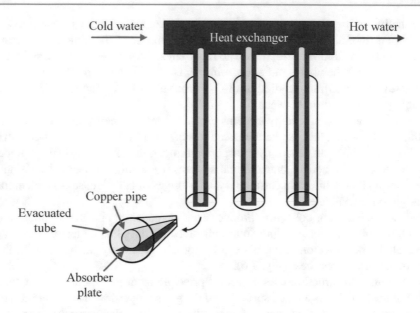

Fig. 7.7 Conceptual schematics of an evacuated tube collector

is usually at higher temperatures than in the flat collectors. The hot internal fluid from many evacuated tubes is collected in a copper manifold where water is heated through a heat exchanger while the internal fluid, usually alcohol, condenses and returns through the manifold. The hot water can also be stored for use at night.

Concentrating Solar Collectors

There are several types of concentrating solar collectors. The concentration ratio of line concentrators is up to 20, while point concentrators produce concentration ratios of up to 1000. Collector configurations are parabolic trough, linear Fresnel lens, central receiver system with dish collection, and central receiver system with the distributed collection.

Parabolic troughs are U-shaped mirrors that concentrate sunlight onto tubes positioned in focus and carrying fluid that can be heated to 400 °C (Fig. 7.8). The fluid is pumped through a heat exchanger to produce superheated steam, which is then used to drive a conventional turbine generator and produce electricity. The solar to electrical energy efficiency in this process is approximately 10%. To obtain maximum solar energy per day, advanced parabolic trough collectors are mounted on one-axis or two-axis sun-tracking devices that point toward the sun.

A parabolic dish collector is a large sun-tracking mirror mounted on a tower. The receiver absorbs the energy and heats the fluid that is pumped through the focal point to 750 °C. Dish collector systems use an array of parabolic dishes with system sizes up to 25 kW and optical efficiency of close to 30%.

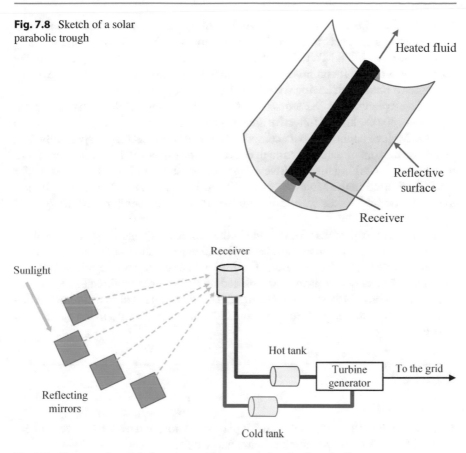

Fig. 7.8 Sketch of a solar parabolic trough

Heated fluid

Reflective surface

Receiver

Receiver

Sunlight

Hot tank

Turbine generator

To the grid

Reflecting mirrors

Cold tank

Fig. 7.9 Conceptual model of a power tower system using distributed reflectors

A central receiver system with distributed reflectors or heliostats can be used to heat a fluid to over 500 °C. The heat is used to generate steam and produce electricity (Fig. 7.9).

There is a commercial solar power plant using 20,000 mirrors (i.e., heliostats) focusing sunlight on a solar power tower filled with salt and heating it to over 500 °C. The salt is then mixed with water to produce superheated steam to drive a turbine generator and produce electricity. This application is suitable for large applications in the MW range.

Case Study 7: Mojave Solar Project The Mojave Solar Project is a 250 MW solar thermal plant located in the Mojave Desert, in California, USA. Operated by Abengoa Solar, the project's construction began in 2011, and the commercial operation commenced in 2014. The plant generates enough electricity to power 75,000 homes and reduces the CO_2 emission by 423,800 tons every year.

This plant helped California achieve its target of 33% of electricity generation from renewables. It was intended for an operational life of 30 years and a yearly production of 617,000 MWh. The plant can store energy as heat for up to 6 h in the absence of sunlight. It comprises two parabolic trough plants, named Mojave Alpha and Mojave Beta, each having a gross peak power of 140 MW.

Mojave's plants have 282 loops or 1128 parabolic trough collectors, encompassing about 780,000 km² of reflective area. A group of loops in a solar field is termed a sector. Mojave Alpha and Beta comprise four and seven sectors, respectively. The plants use mirrors or parabolic trough collectors to collect solar radiation and focus it on heating a working fluid to drive a conventional steam turbine. The heat transfer fluid (HTF) used in the Mojave project is Therminol VP1 oil, which has high thermal stability, low viscosity, and is suitable for high temperature (up to 400 °C) operation.

The Mojave Project was a milestone in the progress of solar thermal technology in the US, but it caused many species to be endangered due to human interference in their natural habitat in the desert. The Mojave Desert contains a total of 29 species and subspecies that have been enlisted as endangered or threatened, such as desert tortoise and Mojave ground squirrel. The project has led to a significant disturbance in the natural ecosystem of many flora and fauna species in the vast area of the desert.

Exercise 7

1. What are the differences between flat plate collectors and evacuated tube collectors? Which one do you think is better and why?
2. Find out the distinctions between line and point solar concentrators.
3. Do some research and gather ideas about the following: solar cooling, solar concentrators, Fresnel reflectors, parabolic dishes with solar Stirling Engines, pumped solar thermal system.
4. A parabolic trough of area 6 m² collects insolation of 1500 W/m² for 6 h. What is the total energy collected by the parabolic trough? [Answer: 195 MJ].
5. A solar thermal power plant has 50 MW capacity and produces 84 GWh of electricity/year. The plant is located in an area with consistent 300 sunny days/year. How many hours per year is the plant producing electricity? [Answer: 1680 h].
6. How much energy in MWh is produced per year from a flat plate collector of area 3.50 m × 3.50 m if the insolation for that location is 4.383 kWh/m²/day? The efficiency of the collector is 38%. [Answer: 7.45 MWh].
7. Discuss the concept of concentrated solar power (CSP).
8. Describe how the Mojave Solar Project in California has disrupted the habitat of natural species. What can be done to ameliorate the disruption caused to natural life? Could you think of any way to avoid this problem for any such large-scale projects in the future?

Solar Photovoltaics

8

Introduction

Solar photovoltaics (PV for short) are solid-state devices that use the properties of semiconductors to convert solar radiation directly into electricity. These devices have no moving parts, generate no noise or emission, and can, in principle, operate for an indefinite time without wearing out. They are modular, reliable, and require minimal maintenance.

A PV system used for remote applications comprises a PV module and other electrical components needed to convert solar energy into electricity and store it until the time of demand. These components include batteries, charge controllers, and inverters as well as other electrical components. The systems can be utility-connected or stand-alone. Some utility-connected systems are large solar power plants.

The areas of the world that can most efficiently use the solar resource and convert solar energy into electricity are those that are exposed to sufficient sunshine. While solar PVs can be used anywhere, they are more effective and less expensive if located in regions with high solar irradiation or insolation.

A large majority of all commercially available solar cells are made from silicon, while a small portion of the market belongs to thin-film solar cells made of a variety of materials.

History of Photovoltaics

Certain metals and other materials such as silicon emit electrons when illuminated by light. This process is termed the photoelectric effect, discovered by Edmond Becquerel in 1839. He observed that the voltage of a battery increases when electrodes were exposed to light.

In 1877, it was reported that selenium exhibits electrical properties when exposed to light. Then, in 1883, the first true solar cell using selenium was invented by Charles Fritts. It was constructed using a thin selenium wafer wrapped with a grid of fine gold wires and a protective layer of glass. The device worked, but it had a poor conversion efficiency of 1%, i.e., the conversion of incident sunlight to electricity was very low.

In the late 1940s, the research on doping semiconductors (materials that can exhibit properties of both an insulator and a conductor) led to the development of a pn-junction and, eventually, the transistor. It was also found that the strong localized electric fields created by the pn-junctions increased the conversion efficiency of photovoltaic cells to about 6%. This first cell in silicon was built in 1954 and was called "solar battery."

The developments continued, and the first practical PV cells were developed for applications in satellites in 1958. The first terrestrial application was for rural telephone systems in the 1950s. With further developments, the cost of early PV cells decreased by thousand times to below $1/W. The exponential growth of solar photovoltaics started around 2010, and it can be attributed to reduced costs of PV materials and large government incentives across the world.

Materials for Photovoltaic Devices

PV cells can be manufactured from a variety of semiconductors, mainly different forms of silicon. Until recently, the majority of solar cells were made from extremely pure monocrystalline Si.

Although monocrystalline Si PV cells are the most efficient commercial cells, they are also the most expensive. Because of that, PV cells are also made from polycrystalline and amorphous Si with a corresponding reduction in conversion efficiency. The less crystalline the material is, the lower is the conversion efficiency of light into electricity. The electrons created by photons (i.e., light) in the cell need to pass through the thickness of the material before being collected at the contacts. The losses that occur during this passage of electrons toward the contacts on the top of the cell increase as the degree of material crystallinity decreases. Besides silicon, solar cells can also be made from other semiconductor materials, such as:

- Compound semiconductor Gallium Arsenide—GaAs. Cells made using GaAs are more efficient than those made using Si, but they are also more expensive.
- Materials that can be deposited in thin layers on flexible substrates: $CuInSe_2$, $CuInGaAs$, and $CdTe$. These compounds are prepared in processes very different from most ubiquitous silicon-based solar cells.
- Polymer solar cells.
- Electrochemical (dye-sensitized) solar cells.
- Nano-crystalline solar cells.
- Hybrid solar cells, i.e., inorganic crystal with polymer matrix.

Thin-film technologies are based on the compound semiconductors such as GaAs, CuInSe$_2$ (CIGS or Copper indium gallium diselenide), CuInSe$_2$ (CIS or copper indium diselenide), and CdTe (cadmium telluride). Modules using CIGS have achieved the highest levels of efficiency with no performance degradation over time. CdTe modules are produced using a simple electroplating process, but they contain Cd, which is toxic.

Each solar cell material absorbs sunlight optimally in a different wavelength range depending on a property called the bandgap. In an atom, the bandgap energy is the difference in energy between the valence band (VB) and the conduction band (CB). It is the minimum energy needed to excite electrons and release them from the atoms of the cell material to create current. Different materials (i.e., elements) have different bandgap energies, based on their atomic structure and nuclear interaction forces (Table 8.1). The bandgap energy corresponds as well to the wavelength of light necessary to excite the electrons from the valence band to the conduction band for selected solar cell materials.

The wavelength shown in the table for each material is the maximum, i.e., the longest photon wavelength (i.e., minimum energy) that can promote an electron into the conduction band. Photons with wavelengths longer than this maximum do not have enough energy to excite electrons. For example, a photon of red light, with $E_{photon} = 1.66$ eV, has enough energy to promote an electron to the conduction band in silicon, gallium arsenide, and cadmium telluride but not in CdSe and CIGS. Silicon solar cells can use all parts of the visible solar spectra but not far-infrared light of a wavelength longer than 1110 nm.

As stated above, a cell can produce electricity only if the energy of photons exceeds the binding energy of electrons in atoms. This value is equivalent to the difference between the two energy levels called the bandgap. Light with frequencies below the cutoff frequency or with energy below the minimum required energy could not produce the photoelectric effect and create an electrical current. Therefore, all photons of light below a certain frequency pass through solar cells without producing electricity. Furthermore, each photon of light excites only one electron, and excess energy does not result in more current but in the generation of heat, causing a temperature increase within the solar cell and performance degradation (high temperature has a negative effect on silicon solar cells).

Because each solar cell material absorbs photons of a specific wavelength more efficiently than the other wavelengths, attempts have been made to construct solar cells that include several different materials layered on each other. These are called

Table 8.1 Properties of selected solar cell materials

Material	Symbol	Band Gap, eV	Wavelength, nm
Silicon	Si	1.12	1110
Gallium arsenide (GaAs)	GaAs	1.42	874
Cadmium telluride (CdTe)	CdTe	1.56	795
Cadmium selenide (CdSe)	CdSe	1.70	730
Copper indium gallium di-selenide	CIGS	1.76	705

multi-junction solar cells and can achieve the highest efficiencies of all solar cells, since more light can be efficiently absorbed and converted to electricity.

Solar radiation can also be concentrated up to 1,000 times using mirrors or lenses. The amount of Si can be reduced by the same factor, which also lowers the cost since mirrors and lenses are cheaper than Si. The efficiency of concentrating PV modules is predicted to double that of conventional systems. The concentrating systems use motors, sensors, and controls to track the Sun's azimuth and elevation to achieve maximum radiation and passive cooling because of the high temperatures.

Principle of Operation of a PV Cell

Because of the dominance of silicon as the solar photovoltaic material of choice, the principle of operation, fabrication, and PV systems will be discussed for silicon-based solar photovoltaics.

The essential structural component of all PV cells is a two-layer configuration, both layers being electrically neutral, but the first layer having a predominantly positive charge (i.e., holes) available for conduction and the second layer having a predominantly negative charge, i.e., electrons) available for conducting the current. Another critical condition for all solar cells is the ability to separate two charges, i.e., electrons and holes before they can collide and recombine. In silicon solar cells, the separation of charge is accomplished by the pn-junction, a created inner structure that features a boundary between the two layers (Fig. 8.1).

The device property described here can be achieved by combining different materials in thin-film solar cells, but it is most easily accomplished in a single material device in silicon because of the extraordinary ability of silicon to have its conductivity modified in a technologically well-controlled process called doping. In this process, impurities are purposely added to silicon in very small concentrations to replace a number of silicon atoms in the crystalline lattice. The whole class of materials called semiconductors, with silicon being the most important, has

Fig. 8.1 Two-dimensional representation of a pn-junction

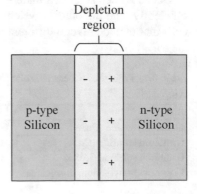

enormous technological importance because of the ability to control their conductivity based on the amount of doping.

With an inner feature in a device, like the pn-junction in silicon, or similar boundaries between two charged regions in other solar cells, the foundation is established for the conversion of sunlight into electricity. When photons of light penetrate a cell and reach across the pn-junction (Fig. 8.1), they can collide with atoms of silicon and excite some of the electrons to a higher energy state.

The light penetrates the cell through the n-type silicon on the top, and the electrons leave the solar cell from the top as well. The conventional current flows from p-type silicon. The significance of the electric field created by the pn-junction is that electrons and holes formed by the light in the vicinity of the pn-junction will be quickly pulled apart in the opposite directions. The electrons and the holes would recombine in the absence of any other effects, and no current would be created. However, due to the presence and effect of an electric field, the electrons are pulled away in one direction (towards the top of the cell), while the holes are forced away in the other direction, and they cannot recombine. In this process, the electron and the hole contribute to the flow of current through an external circuit, and solar power is transformed into electrical power.

Fabrication of Silicon Solar Cells

Fabrication of monocrystalline and polycrystalline silicon solar cells is based on some of the traditional processing capabilities from the semiconductor industry making integrated circuits, i.e., chips. The process starts by purifying the raw material silicon dioxide, SiO_2 (also, sand, silica), in a high-temperature reaction with carbon (Eq. 8.1) and subsequent purification of silicon in a reaction with hydrochloride acid, HCl (Eq. 8.2). The impurities are then removed in a chloride form in a process called fractional distillation and pure (99.99999%), so-called electronic-grade silicon, is recovered (Eq. 8.3).

$$SiO_2 + 2C \rightarrow Si + 2CO \qquad (8.1)$$

$$Si + 3HCl \rightarrow SiHCl_3 + H_2 \qquad (8.2)$$

$$SiHCl_3 + H_2 \rightarrow Si + 3HCl \qquad (8.3)$$

From this stage, the polycrystalline silicon can be processed in two ways to create solar cells (Fig. 8.2). The simpler method is to melt the pieces of polycrystalline silicon (after reaction in Eq. 8.3) in ceramic crucibles, remove after solidification as rectangular blocks, and finally cut into square wafers (typically 10 × 10 cm). Alternatively, from the molten polycrystalline silicon ribbons or sheets of desired thickness can be coated on a thin substrate. These types of solar cells have polycrystalline silicon with so-called grain boundaries that enable recombination of electrons and holes and reduce cell efficiency.

The fabrication of more efficient monocrystalline silicon in the so-called Czochralski process starts by melting the pieces of polycrystalline silicon in a round

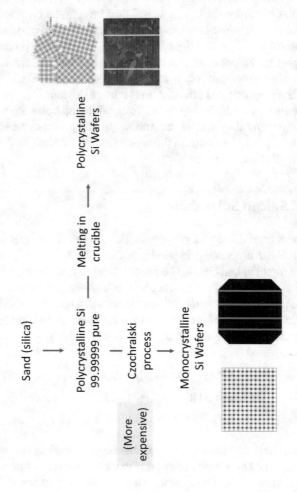

Fig. 8.2 Conceptual diagram of silicon solar cell fabrication and depiction of the crystallinity for two types of silicon

crucible with a lid. A small seed crystal of monocrystalline material is immersed in the melt and slowly pulled out of the melt while the crucible rotates, allowing crystallization, which takes place perfectly replicating the original single crystal planes. After a prolonged 2- to 3-day process, a cylindrical ingot emerges from the melt, and round, approximately 150 μm thick, wafers are cut (i.e., dices).

The diagram of the two processes (Fig. 8.2) also shows the properties of the resulting wafers. The differences between the monocrystalline and the polycrystalline silicon solar cells are depicted in the diagrams of the crystalline structure and images of the two types of wafers. It is visually easy to recognize polycrystalline silicon cells based on the shaded areas and obvious grain boundaries between areas of crystallinity. On the other hand, monocrystalline silicon solar cells are homogenous, and no light variation or reflection is observed. The monocrystalline silicon solar cells are more efficient because their crystalline structure does not contain grain boundaries or other defect sites that enable electron-hole recombination. However, the polycrystalline silicon solar cells have surpassed the monocrystalline cells in terms of the market share because of the lower fabrication cost per unit of power, i.e., W.

Silicon wafers, both polycrystalline and monocrystalline, are usually already doped to make them p-type and further processed using so-called screen-printed cell fabrication. Starting with the p-type wafers, the pn-junction is formed by using a diffusion process to introduce an n-type dopant element and create a pn-junction. After removing the junction from the edges (called isolation), metallic contacts are deposited using the screen-printing method. The contact on the bottom (rear) of a cell is a solid metal contact, while the top contact is patterned, using a screen, to allow light penetration into the cell. Finally, a thin antireflection coating is deposited on the top (or front) of the cell. A cross section of a finished single cell is shown in Fig. 8.3.

Single Cell Operation

When a solar cell is exposed to solar radiation (i.e., sunlight), numerous interactions take place and only a small portion of all the sunlight shining on the cell is used to generate electricity, while the majority of the sunlight participates in reactions

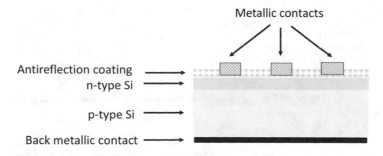

Fig. 8.3 Sketch of the cross-section of a silicon solar cell

without creating a current. The only photons of light successful in generating charge carriers are those that collide with the silicon in the vicinity of the pn-junction and are, therefore, quickly separated by the electrical field before having a chance to recombine.

The interaction between the sunlight and top surface of a solar cell depends on the location where such interactions take place. In general, only a small portion of all the photons or the energy incident on a solar cell is converted to electrons, generating current. The losses include transmission through the cell, reflection from the top metallic contacts or anti-reflection coating, reflection from the bottom contact and photon exiting through the top, recombination in the p-type silicon, and recombination in the n-type silicon (Fig. 8.4).

In a single solar cell, the rear metal contact is connected to the positive terminal of the electric circuit and the metal pads on top of the cell are connected to the negative terminal. These topside metal contacts reflect light and, therefore, must be minimized to leave enough surface area on the top for transmission of light into the cell. The size of the contacts must, however, be sufficient to conduct the current with minimal resistive losses.

The performance of solar PV cells is analyzed using current–voltage curves. The current–voltage or the $I–V$ curve is the basic electrical characteristic of a PV device. It includes all possible current–voltage operating points under specified conditions of incident solar radiation and cell temperature (Fig. 8.5). The basic shape of the curve remains the same for different levels of solar radiation while the short circuit current is directly proportional to the intensity of light.

The two characteristic points on the current–voltage curve are the short circuit current, I_{sc}, at the intersection of the curve with the Y-axis, and the open-circuit voltage, V_{ocv}, at the intersection with the X-axis. The short circuit current is the highest

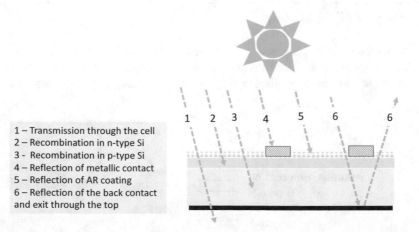

1 – Transmission through the cell
2 – Recombination in n-type Si
3 - Recombination in p-type Si
4 – Reflection of metallic contact
5 – Reflection of AR coating
6 – Reflection of the back contact and exit through the top

Fig. 8.4 Sketch of a cross-section of a silicon solar cell and unsuccessful photon interactions with silicon. The successful interaction in the vicinity of the pn-junction is not shown in the figure

current that can be obtained from a cell at an infinitely small load and at zero voltage. From the short circuit current and as the voltage increases, the current is nearly constant until it reaches a C-shaped section and then sharply drops to the point of maximum voltage or open circuit current at the X-axis.

Additional characteristic points on the curve are obtained by graphing the power ($P = I \times V$) versus voltage and finding current and voltage that correspond to the maximum power (Fig. 8.6).

It is evident from the graph that the voltage, V_{maxP}, is lower when the power is at the maximum than at the open-circuit voltage, V_{OC}, and the current, I_{maxP}, is lower when the power is at maximum than I_{sc}.

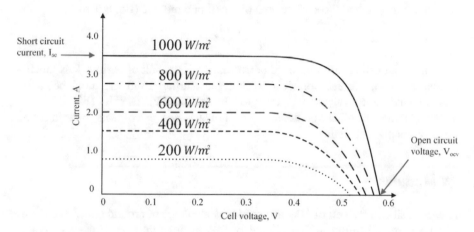

Fig. 8.5 Current–voltage curve for a hypothetical solar PV cell at different solar radiation levels

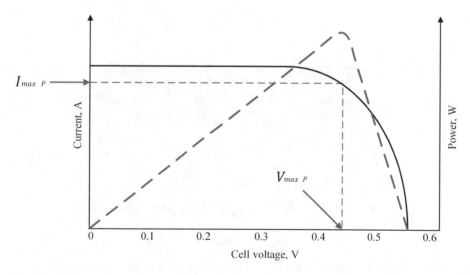

Fig. 8.6 *I–V* curve and power curve for a solar PV cell

The power at the maximum power point is, $P_{maxP} = I_{maxP} \times V_{maxP}$, and it can be compared with the maximum theoretical power which is the product of short circuit current and open-circuit voltage, $P_{max} = I_{sc} \times V_{ocv}$. The ratio between the two is called the Fill Factor (around 80% for most solar cells), and it is used to define the cell efficiency (Eq. 8.4).

$$FF = \frac{V_{maxP} \times I_{maxP}}{V_{OC} \times I_{SC}}. \tag{8.4}$$

The overall energy conversion efficiency of solar photovoltaics (should not be confused with the Fill Factor) is the ratio of the electrical power output and the total solar power input on the cell or module, which is the product of the solar irradiance (i.e., intensity), E, and the surface area of a cell or a module (Eq. 8.5).

$$\eta = \frac{P_{max P}}{E \times A} \tag{8.5}$$

Some solar cells exhibit reduced performance as a result of various loss mechanisms, and their current–voltage curves depart further from the ideal behavior, which would be equivalent to a rectangle having I_{sc} and V_{ocv} as sides (Fig. 8.7). The losses include manufacturer defects, resistance losses in contacts, shading, and other problems with cells.

PV Modules

A typical silicon PV cell of 100 cm^2 produces a voltage of around 0.5 V and a current proportional to the sunlight intensity (3A in full sunlight). However, typical applications require higher voltage and current than one cell can provide. In order to

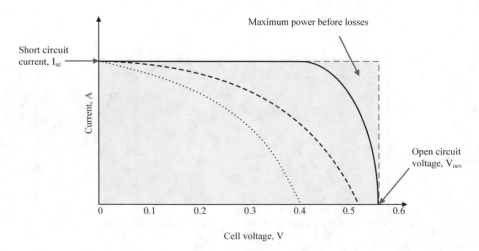

Fig. 8.7 Conceptual representation of a current–voltage curve for a solar cell exhibiting different operational losses

increase the voltage, several cells must be connected in series, and, very rarely, to increase the current, the cells are connected in parallel. In series combination, the positive end of one device or cell is connected to the negative of the other one. The total voltage is the sum of individual cell voltages while the current remains the same as for the single cell. For example, 36 cells in series produce open-circuit voltage of 36×0.5 V ≈ 21.6 V (Fig. 8.8).

A PV module consists of individual solar cells electrically connected to increase the power output. The cells are packaged into modules for electrical, mechanical, optical, and chemical protection using an encapsulant (typically a polymer), a top glass, and a substrate. The modules also have edge sealant and a frame usually made of aluminum (Fig. 8.9).

A module is a stable, long-lasting device with a lifetime of at least 25–30 years. It is constructed to ensure material compatibility and to withstand severe conditions, including extreme temperature variations, ice, snow, hail, wind, UV radiation, air pollutants, and moisture. The encapsulant is the most critical material used in module construction, and the most common encapsulant material, used in commercial PV modules for many decades, has been EVE (ethylene vinyl acetate).

PV Systems

A PV system comprises solar PV modules, batteries, a charge controller, an inverter, switches, cables, and other components (Fig. 8.10). They are typically used to provide power for remote locations without access to the electrical grid, such as navigational tools, telecommunication, cathodic protection, electrical fencing, and water pumping.

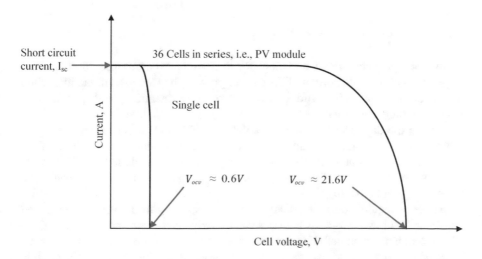

Fig. 8.8 *I–V* curve for 36 solar PV cells in series (typical PV module)

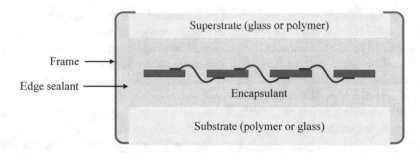

Fig. 8.9 Schematic of a PV module

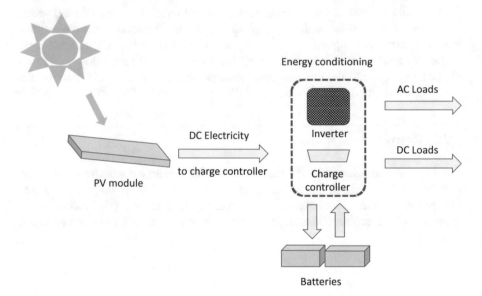

Fig. 8.10 Sketch of a photovoltaic (PV) system

In developing countries, stand-alone PV systems have enormous importance and provide a considerable portion of the overall power. They are used in facilities such as schools or hospitals for lighting, wireless power, refrigeration, water purification, water pumping, and other needs.

During the day, when the sun is available, a PV module is charging batteries, which store electrical energy and release it at the time of demand, usually at night or when the sun is not available. The capacity of batteries included in a PV system depends on the number of days a system must operate even if no sun is available, called "days of autonomy."

Batteries require controlled charging and discharging to deliver their designed capacity. If not used properly, batteries can quickly deteriorate and make the whole PV system non-functional. For this reason, a charge controller is used, which is an electronic device that adjusts the current and voltage coming from the PV module to the battery as well as the current delivered from a battery to load. PV modules produce direct current, and the batteries store and deliver direct current.

The type of batteries most commonly used in remote, off-grid PV systems is a lead-acid battery. It is the longest used commercial battery type. It has modest performance characteristics such as specific energy or cycle life but is generally reliable and low cost. The maximum number of cycles that a lead-acid battery can achieve, before a significant drop in capacity, is around 1000, or for a daily regime of charge–discharge cycling, about 3 years. This makes the batteries the most unreliable component of a PV system, with a lifetime much shorter than for the PV modules. On average, the batteries must be replaced every 3 years.

The lead-acid batteries used in PV systems are different from the most common car starter batteries because of a larger capacity and lower current. Lead-acid batteries are fairly robust to be used in remote applications but need controlled charging and discharging. Another battery type, Li-ion battery with iron phosphate cathode, has been recently introduced to replace lead-acid batteries in PV systems. This battery is characterized by a very high number of cycles, up to 3000, which for PV systems on a daily cycle regime translates to around 10 years.

A charge controller is an electronic device used to control battery charging from the PV module during the day and battery discharging to the loads during the demand. Every battery has a maximum voltage to which it can be charged and a minimum voltage to which it can be discharged. A charge controller is programmed to monitor the battery voltage and disconnect when maximum charging voltage or minimum discharge voltage is reached. A special type of charge controller, called Maximum Power Point Tracking (MPPT) charge controller, has the ability to operate at maximum power point of a PV module, regardless of the battery voltage, and subsequently apply DC-DC conversion and deliver the power to the battery with minimal losses. This is especially important during morning and evening hours when the $I–V$ curve of a PV module is low and maximum power point is shifted; and when the battery is nearly fully charged or fully discharged, requiring low or high charging voltage.

If a PV system is employed to power alternating current loads (AC), an inverter must be added to a system to achieve DC-AC conversion. An inverter is necessary if solar energy is integrated into the electrical power system (i.e., grid-tied system).

Besides these main components, a PV system requires additional, typical electrical parts: wires, switches, fuses, and other components. In addition, remote PV systems have grounding for both the modules and the other electrical equipment.

PV modules can be installed on a roof, or they can be pole-mounted. In both cases, great care must be taken to secure an unobstructed path of sunlight to the surface of the module. Any shading or possibility of debris covering the surface can greatly reduce the performance of a PV system and even cause permanent damage.

Case Study 9: Floating Solar Panels Solar photovoltaic modules require a relatively large land area compared to some other energy generation methods. An alternative is to float PV modules in the water of ponds, lakes, and even hydroelectric water reservoirs. Installing solar PV modules on the hydroelectric water reservoirs creates, in a way, a hybrid renewable energy system. In addition, there is no shading on large hydroelectric dam reservoirs and water provides a cooling effect for the PV modules that improve the performance.

The floating solar modules must have a different construction to prevent water ingress and damage to the cells. They also must be anchored to the water floor to prevent the effects of water currents or waves.

Floating photovoltaics are still in the preliminary stages of their development and their cost is 20–25% higher than for land-based solar installation. However, the amount of power that a well-structured FPV plant can generate is about 10% higher than a similar land-based plant because there are no losses due to shading, and power output is higher because of the cooling effects.

There are floating photovoltaics (FPVs) in more than 35 countries, and their global capacity is estimated at nearly 3 GW. For example, South Korea has a 25 MW FPV plant and is planning a 2.1 GW installation. China already has a 70 MW FPV plant and is developing a 150 MW generation on the Three Gorges dam reservoir. India has plans to build a 600 MW FPV plant. Many countries are exploring FPV plants and the statistics are changing very frequently, indicating that FPV plants might soon contribute a significant portion of the global solar PV energy generation.

Exercise 8

1. Discuss the advantages and disadvantages of solar photovoltaics.
2. Draw the block diagram of a typical solar photovoltaic system that can be used for a residential building.
3. The yellow light given off by a sodium vapor lamp used for street lighting has a frequency of 5.16×10^{14} Hz. What is the wavelength of this radiation in nanometers? (1 nm = 1×10^{-9}m; speed of light in vacuum = 3.00×10^8m/s.) [Answer: 581 nm].
4. A short circuit current for a single silicon solar cell is 0.8 A, and the open-circuit voltage is 0.6 V. The Fill Factor is 0.89. What is the efficiency of this cell in % if the solar irradiance is 800 W/m^2 and the module surface area is 30 cm^2? [Answer: 17.8%].
5. A solar cell material is characterized by the energy for the excitation of electrons to a conduction band of 1.1 eV. What is the maximum wavelength of light in nm (minimum energy) that will generate current in this cell? [Answer: 1100 nm].

Energy Storage

9

Introduction

Renewable energy sources can be classified into two groups in terms of their reliability to deliver power where it is needed at the time of demand. The first group technologies, hydroelectric, biomass, and geothermal energy generation are relatively predictable, and their output can be planned. They are, however, largely dependent on a location and not truly sustainable on a global scale or capable of supplying the entire global power need.

On the other hand, the power from the sun, wind, and ocean has the resource potential many times exceeding the global energy demand. However, these renewable energy sources are all intermittent and unreliable, i.e., they do not deliver power on demand but rather when the sun shines or the wind blows. These energy sources are essentially unreliable unless the energy they generate can be stored and used at the time of demand.

One, often overly simplified argument is that renewable energy can be delivered to the electrical grid and stored there until needed. The problem with this approach is that the electrical grids do not function in such a flexible mode and if there is not another user, who requires power in the relative vicinity, the grid may not be able to "store" the energy from the renewable. The grid may be capable of redirecting the power but not of storing energy. This is also a simplified explanation while bigger issues, e.g., frequency regulation, may be facing grid operators when dealing with the intermittency of the power supply and demand.

It is, therefore, of critical importance for the future of renewable energy sources to enable efficient and inexpensive energy storage methods. The energy generated at any time by the solar, wind, or ocean energy would be, in fact, temporarily stored and then delivered to the electrical grid at the time of demand and would help control many aspects of the electrical grid operation such as frequency and voltage regulation, transient stability services, power factor control, black start, energy time-shift, load leveling, energy arbitrage, and others.

There are numerous methods for storing electrical energy. They include large energy storage systems such as pumped hydro and compressed air, and thermal energy storage and smaller or distributed devices, such as flywheels, supercapacitors, superconducting magnetic energy storage, batteries, and hydrogen. Based on the principle of operation, the energy storage methods are classified as mechanical systems (flywheels and compressed air), electrical systems (supercapacitors and superconducting energy storage (SMES), electrochemical systems (electrolytic capacitors, batteries, and hydrogen/fuel cells), and thermal systems (heat storage and phase change).

Energy Storage Systems

Besides the scale, various energy storage systems are also classified based on the amount of energy they have the ability to store and the power (i.e., current) they can deliver at any time. A generalized comparison shows which energy storage methods offer higher energy and which higher power (Fig. 9.1). There are, of course, nuances to every point of view, and all of the listed technologies can potentially, based on the circumstances of application, be designed to satisfy the requirements across the performance extremes of energy versus power.

Beyond superficial comparison of energy storage capability and power, the technologies are evaluated on many additional factors, including cycle life (i.e., number of cycles without capacity loss), charge time, overcharge tolerance, discharge tolerance, self-discharge, continuous current, operating temperature range, maintenance requirements, environmental impact, and cost. Often, the most quoted performance parameters are specific energy (units, for example, kWh/kg) and energy density (units kWh/L), or specific power and power density (kW/kg and kW/L).

However, these criteria are more important for different application segments such as portable or transportation and less applicable for stationary-type applications in renewable energy systems. The exemption to this can be found in the futuristic concept of massive charging of electric vehicles using renewable electricity generation and storing the energy in the vehicle batteries, only to be returned to the grid at the time of demand. Despite being fairly inadequate for evaluating the best

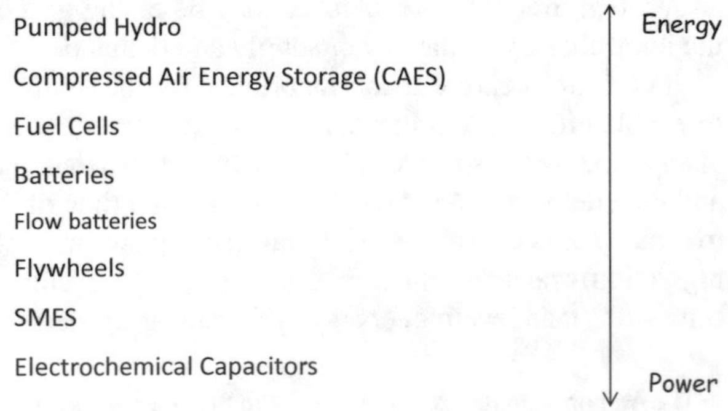

Fig. 9.1 Generalized comparison of energy storage methods

Pumped Hydro   ↑ Energy

Compressed Air Energy Storage (CAES)

Fuel Cells

Batteries

Flow batteries

Flywheels

SMES

Electrochemical Capacitors   ↓ Power energy storage methods for renewable energy sources, the Ragone plots of power density versus energy density give nonetheless a useful overview of the energy storage methods and point out that the best devices overall would be expected in the upper right corner (Fig. 9.2). Note that in the Ragone-type graph, the power, and energy density are not necessarily connected nor that the properties influence one another but rather that they are plotted together for a convenient demonstration.

Technologies of interest for renewable energy systems include in particular pumped hydro storage (PHS), compressed air energy storage (CAES), and electrochemical energy storage devices (batteries, fuel cells, and supercapacitors). It is evident from the figure that batteries have good energy or discharge time but lower power. Various battery types cover different ranges of power and energy. The flywheels and supercapacitors have high specific power despite the low specific energy, meaning that they can deliver high current for short periods of time. Batteries and fuel cells, on the other hand, have the lesser current delivery capability but higher specific energy, which means that they can run for a long time.

Pumped hydro energy storage utilizes the potential energy of water, lifted to a height, to generate power in a hydropower turbine and generator. The energy, when available in the excess of consumption, is used to power a water pump and lift water from a lower to an upper water reservoir, store the potential energy in water, and release through a special turbine, that rotates in both directions, to drive a generator and generate electricity. Water is typically pumped to the upper reservoir during the night when there is excess power available, e.g., wind power is larger at night in some areas, and released through a turbine during the day to produce electricity.

This is currently the most effective large energy storage method, with about 80% of electrical energy recovered. Some losses occur because of evaporation and pump/

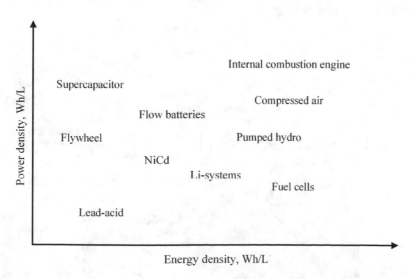

Fig. 9.2 Comparison of power and energy density for selected energy storage devices. The comparison is not based on specific data but rather on conceptual, averaged, or generalized performance

generator inefficiency. The pumped hydro energy storage is the most effective for large-scale electricity storage and generation (e.g., >2000 MW). The technology is dependent on the location and geography, and it comes at huge capital costs. It has similar environmental effects as hydropower generation, including blocking natural water flows, disrupting the aquatic ecosystems, flooding previously dry terrain and changing landscape, and potentially destroying or displacing terrestrial wildlife. In addition, pumped hydro water dams may trap and kill fish and have further ecological effects that are usually mitigated with appropriate actions.

Compressed air energy storage systems (CAES) use off-peak electricity to compress air and store it in reservoirs, either underground caverns or aboveground pipes or vessels. When electricity is needed, the compressed air is released, heated by combusting natural gas, which expands the air and drives a turbine (Fig. 9.3).

The CAES generation has about 50% efficiency, high specific power of up to 2 MW/kg, and fast start-up. However, it depends heavily on the location and geological structure. Above the ground facilities are typically smaller than underground, can generate up to 50 MW, and operate from 2 to 6 h. Underground facilities are considered more cost-effective, they generate up to 600 MW. They rely on structures such as limestone caverns or unused salt mines.

CAES facilities require massive construction and are only economical on a very large scale. The first commercial CAES plant in the world was opened in Huntorf (Germany) in 1978. It has a capacity of 290 MW of power. The air is compressed in two storage caverns holding 10.6 million ft^3 (0.3 million m^3). The generator ramps at a fast rate of 90 MW/min to reach the full load. The first commercial CAES plant in the United States, the McIntosh facility, was opened in 1991. It has a power rating of 110 MW, and the storage is in a single cavern of salt-dome type, holding 19.8 million ft^3 (0.56 million m^3). The power plant is capable of delivering 2600 MWh of electricity without recharging.

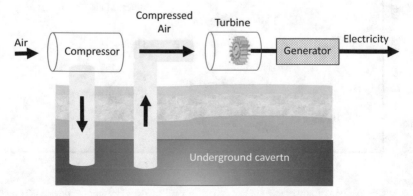

Fig. 9.3 Sketch of an underground compressed air facility with electricity generation

Electrochemical Energy Storage

Batteries, fuel cells, and electrolytic capacitors are all based on the electrochemical principles of operation. All electrochemical energy storage devices have unique properties, suitable for certain applications, and capable of providing energy storage in combination with renewable energy generation.

Batteries

Batteries contain active masses on the electrodes, which change during charge and discharge. The capacity or the amount of energy stored in batteries depends on the amount of the active mass and its properties. These reactions are reversible (not in a strict thermodynamic sense) in rechargeable batteries (i.e., secondary batteries) and irreversible in primary batteries. The chemical reactions that occur in rechargeable batteries during discharge (i.e., while delivering energy) are reversed when the electricity is applied to the battery during charging.

When selecting a battery for large-scale electrical energy storage or distributed renewable energy system, numerous factors have to be considered. Some of the most important factors are the energy density (capacity per unit weight or volume), load current, single-cell voltage, self-discharge rate, charging time, cycle life, shelf life, temperature impact, environmental impact, and cost. It should be obvious that these requirements carry different importance for different applications. For example, energy density, i.e., the amount of energy stored per unit volume, is critically important for portable (e.g., cell phone or laptop) and automotive applications but may seem less relevant for stationary applications in conjunction with renewable energy generation. Or, the load current is very important for power tools where a high current, i.e., power, is needed, but capacity or the length of time the battery performs is not so critical. For some remote renewable energy systems, the most important factor may be the cycle life. This is also true for stationary applications of batteries where energy density is not so critical, but cycle life and cost are the most important. The price of a battery is determined based on how much energy it can deliver.

There are four major battery systems in use: lead-acid (LA), lithium ion (Li-ion), nickel metal hydride (NiMH), and nickel cadmium (NiCd). They are characterized by different performance merits and properties (Table 9.1).

The comparison of the most important batteries based on their energy density is shown in Table 9.2.

Batteries are compared on the basis of a number of performance factors. The state of health (SoH) is used to determine the level of battery degradation from 100% when manufactured. The state of charge (SoC) compares the level of available capacity at any point with the full capacity, where 100% means a fully charged battery and 0% a fully discharged battery. Depth of discharge (DoD) is the inverse of the state of charge. Charge or discharge rate is given relative to its capacity and expressed as so-called C-rate, which refers to the time required to fully charge or

Table 9.1 Characteristics of the four major types of batteries

Battery type	Voltage, V	Anode	Cathode	Electrolyte
Lead acid (LA)	2.0	Pb	Lead oxide	Aqueous sulfuric acid
Nickel-cadmium (NiCd)	1.2	Cd	Nickel oxide hydroxide	Aqueous potassium hydroxide
Nickel-metal hydride (NiMH)	1.2	Metal Hydride	Nickel oxide hydroxide	Aqueous potassium hydroxide
Lithium Ion (Li-ion)	4.0	Li (C)	Lithium cobalt oxide	LiPF$_6$

Table 9.2 Comparison of energy densities for the main battery systems

Battery type	Energy Density (MJ/kg)	Energy Density (Wh/kg)
Lead-acid	0.13	35
NiCd	0.16	45
NiMH	0.28	90
Li-Ion	0.54	150
Gasoline	43	12,000

discharge a battery. Perhaps the most important property for renewable energy applications is the cycle life or the number of full charge-discharge cycles a battery can undergo without reduction in capacity. For many applications, capacity reduction below 80%, compared with the initial capacity, is considered the end of life. Typical batteries last 300–3000 cycles.

Lead-acid (LA) batteries were the first commercially successful batteries. They are built using one lead and one lead dioxide electrode immersed in sulfuric acid (H_2SO_4). The equation for the reactions in lead-acid batteries is shown in Eq. 9.1.

$$Pb + PbO_2 + 2H_2SO_4 \leftrightarrow 2PbSO_4 + 2H_2O \qquad (9.1)$$

The reaction proceeds from left to right during the discharge of a battery and from right to left during the charging of the battery. It can be seen from the equation that the only product of the discharge reaction is lead sulfate ($PbSO_4$) on both electrodes. Lead sulfate has a tendency to grow in crystal size and become unusable. The consequence of this is that the LA battery cannot be left in the discharged state for very long. These batteries have the lowest energy density of all four major battery types, but they are cheap and rugged. They still hold roughly 50% of the battery market, primarily due to their use as starter batteries for vehicles.

Nickel Cadmium (NiCd) batteries were first commercialized in the mid-twentieth century and have remained one of the most durable batteries on the market. They have better energy density, i.e., higher capacity per unit weight (45–65 Wh/kg) than lead-acid batteries, and superior performance compared to all other batteries in terms of load current, a higher number of cycles, and low-temperature performance. Elemental cadmium on the anode is oxidized during the discharge reaction, while nickel oxyhydroxide is reduced on the cathode, and oxides are produced on both electrodes (Eq. 9.2).

$$2NiOOH + Cd + 2H_2O \rightarrow Cd(OH)_2 + 2Ni(OH)_2 \qquad (9.2)$$

NiCd batteries are often labeled as problematic due to the "memory effect," a phenomenon whereby some of the active mass becomes ineffective if a battery is repeatedly discharged only to a specific level less than a full discharge. This, however, is not a serious problem if an NiCd battery is properly "maintained" and fully discharged at least once a month. The more serious problem for NiCd batteries is an increased awareness of cadmium toxicity and the resultant regulations in many countries to discontinue the use of these batteries.

Nickel Metal Hydride (NiMH) batteries represent a variation in NiCd batteries. In fact, the nickel (more precisely NiOOH) electrode is common for both battery systems. The other electrode, instead of the cadmium electrode, is made of special compounds with the ability to hold hydrogen. These materials are called metal hydrides. NiMH batteries have better energy density than NiCd batteries, but lower current capability and lower number of cycles. High energy density had made these batteries a popular choice for portable electronics and electric vehicles in the 1990s.

Lithium-ion (Li-ion) is the most advanced battery type based on its high specific energy. There are several versions of lithium-ion batteries, and their main common feature is the use of Li-ion as the negative electrode. Lithium is the metal with the strongest tendency to be oxidized and, therefore, the most promising negative electrode material. The positive electrode composition varies, and the most common positive electrodes are (lithium) cobalt oxide, (lithium) manganese oxide, and (lithium) iron phosphate. Li-ion batteries show the most superior energy density of all batteries, i.e., they can deliver several times more capacity than other batteries in the same volume and weight. They are, however, the most expensive batteries, very sensitive to improper charging, and prone to accidents resulting in explosion and fire. The safety has been greatly improved since the use of pure Li electrodes at the beginning of the 1990s, but some concerns still remain, and Li-ion battery air transport is strictly controlled.

The lithium-ion battery reaction mechanism is based on the anode and cathode host materials and Li^+-conducting electrolyte. In the anode, Li is contained as ions in the crystals of graphite. During the discharge, Li^+ travels through the organic electrolyte to the cathode and merges (i.e., intercalates) there with the cathode structure, most commonly cobalt oxide or iron phosphate. During the charging stage, the process is reversed. Li-ion batteries have been the focus of intense research, and they represent the best opportunity for finding a battery system that would satisfy the needs of energy storage for a variety of applications. The research concentrates on finding the lightest but most conductive materials for both negative and positive electrodes. One particular variation widely investigated is the lithium-air battery, where the positive electrode does not hold any active mass and instead draws oxygen from air for the reaction.

There are many other types of batteries, such as flow batteries, Li-polymer batteries, sodium beta batteries, metal-air batteries, and many others. (For further reading consult reference 1).

Hydrogen and Fuel Cells

Another energy storage method is based on using hydrogen as the storage medium and using it in fuel cells to produce electricity. Because hydrogen is the substance with the highest energy content using it as a fuel in fuel cells is efficient in terms of energy density. The biggest challenge is that hydrogen is not freely available in nature and must be first extracted, i.e., released, from the bound state in other substances, such as water or fossil fuels. These processes are endothermic, i.e., they require energy. The paradox of using hydrogen as a fuel is that energy must be first spent to produce hydrogen only to gain less energy back when hydrogen is used to generate electricity.

Further consideration of this idea leads to the conclusion that the only way this concept would make sense is if fuel or energy from which hydrogen is produced comes from renewable energy sources, and an obvious example of such a hydrogen-based economy is the specific concept of a solar-hydrogen cycle. The solar-hydrogen cycle starts with water electrolysis to produce hydrogen, followed by hydrogen storage, and distribution and is completed when hydrogen is used in fuel cells to produce electricity at the time of demand. The electrical energy used in electrolysis is supplied from a renewable energy source such as solar photovoltaics, wind energy, hydroelectric energy, ocean energy, or geothermal energy. Hydrogen can also be directly obtained from fossil fuels or biomass.

When hydrogen is produced in the electrolysis reaction the cycle starts and ends with clean water, so there is no emission of any kind or imbalance in nature as a result of energy production. After the storage, hydrogen can be distributed to the point of use through a network of pipelines. Effectively hydrogen becomes an energy carrier or energy vector.

In general, there are four stages in the hydrogen economy concept, namely, production, storage, distribution, and use as fuel. Hydrogen can be obtained from the electrolysis of water for which electricity from solar, wind, or another renewable energy source is used. Electrolysis is also an electrochemical process and requires two conductive electrodes immersed in an electrolyte (e.g., acid, base, salt). When the voltage is applied to the electrodes, water is decomposed into hydrogen and oxygen (Eq. 9.3).

$$H_2O + \text{Electricity} \rightarrow H_2 + \frac{1}{2}O_2 \qquad (9.3)$$

The voltage applied across the electrodes must be sufficient to overcome the bond strength between the hydrogen and the oxygen atoms in water. The minimum value, i.e., the theoretical decomposition voltage is 1.23 V at 25 °C and 1 bar pressure, while the practical voltages in water electrolysis reactors (i.e., electrolyzers) exceed 1.8 V. Water is a poor conductor of ions and requires a conductive electrolyte for sufficient cell voltages. The most common electrolytes are potassium hydroxide (KOH), sulfuric acid (H_2SO_4), and a proton-conducting solid polymer membrane.

Perhaps the most ideal way of producing hydrogen would be a photoelectrochemical process in which solar radiation would supply energy to electrolysis cell, and hydrogen would be produced. This method, which contains elements of both solar photovoltaic electricity generation and water electrolysis, has been the subject of intense research efforts. The practical applications of this technology, however, still remain elusive.

As an energy source, hydrogen can be stored as a compressed gas, liquified hydrogen, in form of metal hydrides, complex hydrides, or in methane, methanol, ethanol, and ammonia, and extracted easily before use.

A fuel cell is a device for the electrochemical conversion of chemical energy into electricity without any intermediate steps. Fuel cells operate as long as there is a continuous supply of fuel and oxidant. The most common fuel is gaseous hydrogen, but other fuels such as methanol, methane (natural gas), and others can be used as well. The fuel and an oxidant are continuously supplied to electrodes, which are separated by a membrane or an electrolyte. Fuel reacts on the anode and is oxidized in the reaction while generating electrons, which then flow to the other electrode through an external circuit and loads, to produce power. The hydrogen ions, i.e., protons travel through the electrolyte to the cathode where they react with oxygen and electrons to form water, as the only product of reaction and energy. The concept of a fuel cell with acidic electrolyte is demonstrated in Fig. 9.4. The most commonly used fuel cell, polymer electrolyte membrane, uses acidic electrolyte in form of a thin polymer film.

The overall reaction in a fuel cell is shown in the Eq. (9.4).

$$O_2 + 2H_2 \rightarrow 2H_2O \tag{9.4}$$

The only by-product of a hydrogen fuel cell reaction is water. The particular fuel used in the fuel cell determines its output voltage. The theoretical output voltage of a hydrogen fuel cell is 1.23 V, although practical values are typically around 0.7 V. There are several reasons for voltage losses in all fuel cells. These include slowness of the reactions on the electrodes, conductivity losses, and losses due to reduced concentration of reactants in the reaction zone. The voltage of every fuel

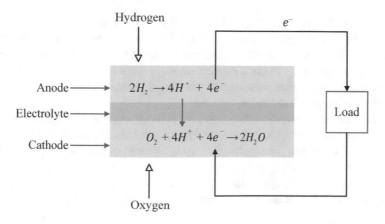

Fig. 9.4 Sketch of a fuel cell with acidic electrolyte

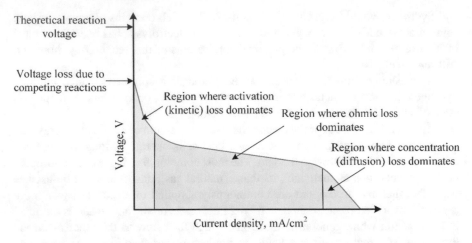

Fig. 9.5 The voltage–current relationship for a fuel cell, including sources of losses

cell changes with the current drawn from it or load. A typical voltage-current curve is shown in Fig. 9.5.

The efficiency of a fuel cell is expressed as a ratio between the practical cell voltage and the theoretical voltage (Eq. 9.5).

$$\varepsilon = \frac{E_{cell}}{E^O} \times 100\% \tag{9.5}$$

The efficiency of a fuel cell is often compared to that of combustion reactions, which are limited by the Carnot Cycle. For a combustion engine, the maximum theoretic efficiency depends on maximum and minimum temperature. The Carnot efficiency is calculated simply using these two temperatures (Eq. 9.6).

$$\eta_{Carnot} = \frac{T_{hot} - T_{cold}}{T_{hot}} \times 100\% \tag{9.6}$$

This is the theoretical maximum efficiency of a combustion engine while the actual or practical efficiency is usually much lower. In a combustion engine, chemical energy is initially converted into thermal energy, then to mechanical energy, and eventually to electricity. In each conversion step, there is an energy loss, reducing the overall efficiency. Fuel cells, on the other hand, directly convert the fuel's chemical energy into electrical energy.

Besides the advantages in efficiency, fuel cells are emission-free if hydrogen is a fuel, they generate no noise, and are extremely durable with lifetimes in decades.

A fuel cell describes the whole category of devices that utilize an electrochemical system to continuously generate electricity from a fuel. From a technological and manufacturing perspective, fuel cells differ considerably and are usually distinguished based on electrolyte and operating temperature (Table 9.3).

Table 9.3 Overview of fuel cell types

Fuel cell type	Electrolyte used	Efficiency, %	Operating temperature, °C
Alkaline fuel cell (AFC)	35–50% KOH	50–60	60–90
Proton exchange membrane fuel cell (PEMFC)	Polymer membrane, such as Nafion	50–60	50–80
Phosphoric acid fuel cell (PAFC)	Concentrated H_3PO_4	50–55	160–200
Molten carbonate fuel cell (MCFC)	Li_2CO_3/Na_2CO_3	60–65	620–660
Solid oxide fuel cell (SOFC)	ZrO_2/Yt_2O_3	55–65	800–1000

Electrolytic Capacitors

Electrolytic capacitors are based on electrodes with a high surface area that accept a large amount of charge and can deliver high currents during fast discharge. The electrolytic supercapacitors do not involve any active mass that chemically changes during charge and discharge and therefore they have a very small amount of capacity.

Electrolytic supercapacitors are electrochemical devices, but they do not rely on chemical reactions for the generation of electricity. Rather, as their name says, they are equivalent to capacitors. The energy is stored as charge on large surface area electrodes. The two electrodes contain the opposite charge, and they are spaced very close to each other with an electrolyte in between. These are also called electric double-layer capacitors (EDLC), as a double electrochemical layer is always created at the interface between electrode and electrolyte. The commonly used electrodes are made from activated carbon, while the electrolyte is typically KOH, organic solutions, or sulfuric acid. The key development principles are to increase the surface area of electrodes in direct contact with the electrolyte and to decrease the distance between the electrodes. The capacitance of a supercapacitor (in Farads) can be calculated from the surface area of electrodes, A, distance between the electrodes, d, and electrical permittivity, ε (Eq. 9.7).

$$C = \frac{\varepsilon A}{d} \tag{9.7}$$

Case Study: Compressed Air Energy Storage Projects CAES is one of the least explored technologies for ES, with only two operational CAES plants in the world, in Huntorf, Germany, and Alabama, USA. Both plants use underground salt caverns for the storage of compressed air.

In 2013, a group of researchers from the Bonneville Power Administration (BPA) and the U.S. Department of Energy's Pacific Northwest National Laboratory (PNNL) explored the potential of two sites in eastern Washington for CAES plants in the Pacific Northwest, the Columbia Hills site, and the Yakima Minerals site. The study examined whether natural, porous rock reservoirs could be used to store compressed air and the geothermal energy to heat and expand the air to drive a turbine. Electricity from geothermal heat would also be used to power the compressors. The abundant wind power blowing at night in the region could be stored as compressed air and then released when necessary.

The Columbia Hills Site has the potential for power generation of 207 MW. The levelized cost of electricity from this site was estimated to be only 6.4 ¢/kWh. The site is close to the natural gas (NG) pipeline, making it easy to utilize the gas to expand the air to generate electricity.

The Yakima Minerals site underground reservoir is over 10,000 ft. deep. The site has a power generation capacity of 83 MW. The total capital cost for this project has been estimated at $2738/kWh. The idea of the two CAES plants was to incorporate renewable power and energy storage together in order to optimize both technologies.

Exercise 9

1. Explain the necessity of installing an energy storage system with a residential solar panel.
2. Describe the five main types of fuel cells with relevant chemical equations and diagrams.
3. If two or more types of energy storage systems are combined, it is termed a hybrid energy storage system. Find out some practical examples of hybrid energy storage systems along with the type and size/rating of the storage system used.
4. Discuss some grid applications of energy storage, such as load leveling, energy arbitrage, and frequency regulation.
5. The upper reservoir of a pumped storage system has a surface area of 8 km^2 and lies 370 m above the lower reservoir. How much energy is stored if overnight pumping raises the water level in the top reservoir by 1 m? [Hint: Potential energy = density × volume × g × height] [Answer: 8×10^6 kWh]

Reference

1. S. Petrovic, "Battery technology crash course", Springer, New York 2020, eBook ISBN 978-3-030-57269-3; Hardcover ISBN 978-3-030-57268-6

Grid Integration of Renewable Energy 10

Introduction

Most of the conventional electricity grids are powered by coal or gas-fired power plants. Generating electricity using different renewable energy sources (RESs) such as wind, hydro, solar, geothermal, and biomass is gaining popularity due to growing concerns about the environment and the imminent depletion of fossil fuels.

Grid integration of renewable energy implies the supply of renewable electricity to run utility grids. This involves developing cost-effective and efficient methods to ensure the grid's stability and reliability despite the intermittent and fluctuating nature of renewable sources.

The purpose of the utility grid is to reliably supply readily available power with a constant voltage and frequency. Integrating RESs into the utility grid offers some challenges and is not a smooth process. When the concepts of power transmission and distribution systems emerged, there was no awareness of renewable energy sources. As a result, the electricity grids were not designed to comply with RESs. The electricity grid is expected to supply a linear power output with reliable power quality. Power generators running on conventional sources of energy can be controlled easily. Grid operators have the control to start, stop, or vary the power generation after a definite time constant by varying the fuel supply to energize or de-energizing the prime mover.

Grid integration of RESs is essential, but it has to be done correctly so that the advantages of renewable integration are accentuated higher than the associated inconveniences. Most electricity grids use AC power transmission, but renewable electricity is sometimes DC in nature. This necessitates conversion from DC to AC power. The power translators must maintain a fixed voltage and frequency and ensure power quality and stability by mitigating the intermittency of renewables. This helps independent power producers to participate in energy markets and businesses. Even if a RES produces AC power, e.g., from a wind turbine, it must be

E. Hossain, S. Petrovic, *Renewable Energy Crash Course*,
https://doi.org/10.1007/978-3-030-70049-2_10

coupled through a power converter to ensure that it has the correct voltage and frequency level to synchronize the output power with that of the utility grid.

One disadvantage of RESs is their intermittent nature—a constant amount of power is not continuously available. Storage devices are often used along with renewable power generators to mitigate the intermittency. Biofuels, hydroelectric power, and some forms of geothermal power are reliable power sources, but these also must be coupled through a power translating device to ensure power quality.

Another negative impact of grid integration of RESs is the creation of the "duck curve" in the power demand. Figure 10.1 shows an example of the "duck curve" using the load demand curve in the presence of solar power generation. The grid power demand declines after midnight and ramps up after sunrise. Around 9 am, the sun shines brightly, producing a lot of solar power, thereby cutting down the demand from the grid since the power is generated from the solar panel.

The sunlight intensity increases, proportionally increasing the production of solar power. At this time, the grid may have overgeneration, which is a burden to the transmission and distribution system and causes misuse of resources. As dusk arrives, solar power drastically falls, which stresses the utility grid because of a sharp increase in demand. Then grid demand falls again during the night. Every year, the amount of electricity production from RESs increases, making the duck curve steeper and making its shape more evident. This is perhaps the strongest reason for the reluctance to grid integration by the grid operators.

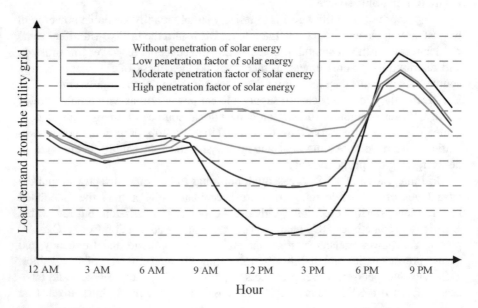

Fig. 10.1 Illustration of the duck curve. Renewable power sources such as solar cause risk of overgeneration and necessitate fast ramping up of power—two undesirable situations for the power system

Global Scenario of Grid Integrated Renewables

In the United States, independent producers of renewable electricity are allowed to send into the utility grid as much renewable electricity as they can, after meeting their consumption requirements.

The grid operators also maintain the continuous availability of electricity and the power quality. There is a good backup plan to supply power from neighboring states if any major flaw in the system is caused by any independent renewable power source. The large geography and developed infrastructure of the utility grid in the United States allow flexibility and ease of grid integration of RESs, which is necessary to encourage people to increase power generation from renewables and foster this growing technology.

In Europe, the picture is quite different since the grid has significantly matured in terms of RESs, and there are strict rules to abide by to integrate RESs into the utility grid. Independent renewable power producers in Europe are only allowed to send a certain amount of power into the grid at a certain period of the day. The power level and power quality must be maintained accurately by the suppliers, and the intermittency of the supply must be addressed.

In 2019, roughly 15% or 64 million MWh of the total electricity produced in the United States was sourced from RESs, of the total 411 million MWh of electric energy generated. Canada produces 67% of its electricity from RESs, primarily from hydropower. Bloomberg New Energy Finance (BNEF) predicts that by 2040, Europe will have 90% of its electricity generation from renewables, and solar and wind alone will contribute 80%. As of September 2020, Australia has reached a milestone of over 50% of its electricity from RESs, with 24.6% from rooftop solar, 10.1% from large-scale solar, 13.5% from wind, 1.95% from hydro, and 0.2% from biomass. By 2025, Australia could attain around 75% of its electricity from RESs. China was generating 26.7% of its power from RESs as of 2018.

Principles of Grid Integration of Renewables

Renewable energy sources may be integrated into the utility grid at all voltage levels, ranging from distribution-level low voltages to transmission-level high and extra high voltages. The size and rating of the different modules used at different levels are different, but the basic principle of a grid-tied RE system remains more or less the same. When integrated into the grid, RESs such as solar PV can perform two functions—either supply all of its produced power to the grid or meet the local loads and then supply the remaining power to the grid.

Since RESs fluctuate in nature, an energy storage system (ESS) is necessary to store excess energy during overproduction and discharge it when there is a high demand but no RE production. For example, solar energy can be stored in a battery during the day to supply the loads at night. ESSs can be used in both rooftop or small residential RE systems and large-scale utility RE systems. The solar PV arrays produce DC power, which needs to be converted into AC power of a definite voltage level (120 V or 220 V) and a constant frequency (50 or 60 Hz ± 2%) before grid integration. This needs to be accomplished using a DC/AC inverter. Grid synchronization is essential for any power generating source during connection to the utility grid. Therefore, after DC/AC conversion, solar power also needs to be synchronized to the grid power. Filters, capacitors, circuit breakers, etc., are also necessary to filter out the DC components and ripples from the power output and ensure necessary electrical protection.

A RE-sourced power generating system requires some very essential equipment irrespective of it being grid-tied or stand-alone. The additional equipment is together referred to as the balance of system (BoS), which involves a large share of the investment or capital costs. Several pieces of equipment are involved in a RE system, such as energy storage or batteries, charge controller, power-conditioning equipment (PCE) or inverter, safety and protection devices, filters, meters, transformers, etc.

Types of Inverters used for Grid Integration

Inverters are extremely important in the study of grid integration of RESs and deserve special attention. Typically, four-quadrant inverters are used for grid integration, since these inverters are capable of bidirectional power flow and supplying or absorbing both real and reactive power. The supply or absorption of reactive power is a critical concern because it impacts the power factor of the utility grid and has a direct impact on the power quality and the billing procedure. The inverters for grid integration can be battery-less or battery-based. Inverters may also be classified based on their connection with the generating system, such as solar panels and the grid.

Grid-tied battery-based inverters are multifunctional, flexible, adaptive, and can perform net metering and sell energy to the grid. Battery-less grid-tied inverters simply convert the solar DC output into AC power, feed the loads, and send the surplus to the grid without any storage. These can also perform net metering services, are the least expensive, and the simplest. Hybrid inverters are advanced inverters that can control the solar panels through Maximum PowerPoint Tracking (MPPT) algorithms and the charging and discharging of the batteries through charge-controlling functions.

All these constitute a complete package designed for backup power and net metering services. They are the most advanced and expensive but can manage the transfer of power from the solar panels to the battery, load, and grid, and also from the grid to the battery and the load.

Based on the connections, inverters can be classified into four types: central, string, multi-string, and module integrated inverters (Fig. 10.2). If all the individual generating units are connected to a single inverter, the arrangement is known as a single or central inverter. This is used in mid- to large-scale projects. The modules in a string are in series, and the strings are in parallel to the inverter.

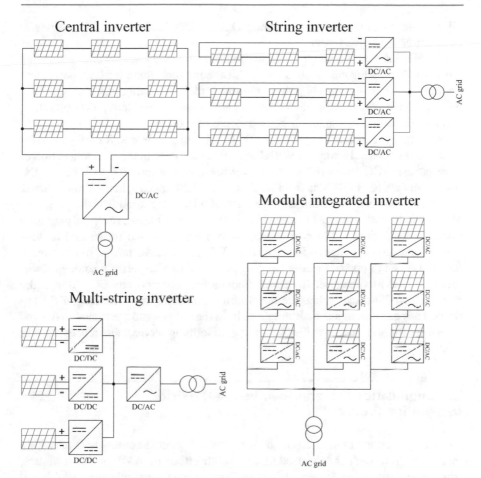

Fig. 10.2 Different types of inverter configurations, central or single inverter, string inverter, module-integrated inverter, and multi-string inverter

The input of such an inverter is a common DC bus, and the output is a common AC bus. This configuration is simple, cheap, and is well suited for open places with uniform insolation, inclination, and shading.

String inverters can connect a string or a series connection of several generating units or modules together. Several string inverters are then connected to a common AC bus. This setup has better efficiency than a central inverter, but it is risky because the overall power might be reduced if one generating unit faces issues, such as shading. This configuration is used in large solar fields with varying levels of inclination and insolation.

A multi-string inverter is a combination of the previous two configurations. This topology uses a DC–DC converter with every string of generating units. The converter converts the output into another DC value. The DC values from the output of all the converters are fed to a common DC bus that feeds a DC–AC inverter to finally produce an AC power.

Module integrated inverter configuration constitutes a microinverter integrated with each generating unit or solar panel. All the microinverters feed AC power to a common AC bus that injects power into the grid. This configuration is relatively expensive and very reliable because each panel is an independent power producer, and the production of one unit is not affected by the others.

Solar panels generate DC power, so DC/AC inverters are used for grid-tied solar generating systems. However, DC/DC converters are used in the interconnection of solar panels to DC microgrids or DC power lines, such as the 3100 MW Pacific DC Intertie or Path 65 HVDC line from Washington to Los Angeles, in the United States.

There is no need for inverters in the case of wind integration into the grid, provided that the power is synchronized to the grid by other means. Inverters play various roles other than power conversion, so they are employed for several reasons even if they are not essential for converting DC power to AC power. In addition to the conversion of power, inverters in grid-tied systems also perform some grid services. In case of voltage or frequency fluctuations, inverters can ride through the minor disruptions and isolate the generating unit during more prolonged disruptions. They can also adjust their output to hold the voltage and frequency to desired levels. Inverters can perform black start or grid forming services and reactive power control.

Communication Systems Involved in the Grid Integration Process

Several systems are used to ensure continuous and secured communication within electricity grids, such as Advanced Metering Infrastructure (AMI) or Smart Meters, Wide Area Monitoring System (WAMS), Power Line Communication (PLC), and Energy Management Systems (EMS). A hybrid of several technologies involving fiber optics, copper wires, and wireless technologies is also possible.

In the case of communication technologies, some important considerations need to be made, such as security, reliability, bandwidth, scalability, latency, and costs. A communication system adopted in typical electrical grids comprises a high bandwidth backbone, which connects several low bandwidth access networks. For the backbone network, fiber optic cables or digital microwave radio waves may be used, and for the access networks, twisted pair cables, PLC, or wireless systems may be used.

Control Systems in the Grid Integration Process

To maintain proper grid synchronization, a robust monitoring and control system is crucial. Control is essential to operate the inverters, which decide the direction of active or reactive power flow to or from the utility grid and storage devices. Control is also required to switch from grid-tied to "islanded" modes and vice versa. In order to maintain grid stability and a constant voltage and frequency level, control is indispensable.

Automatic Generation Control (AGC) systems are used to maintain the reliability and stability of the power systems and adjust the output of the generating units responding to real-time load variations within seconds to minutes. The control systems might be centralized or decentralized. Communication is an inherent part of the control, and so, the communication and control in grids are often handled together by using the Supervisory Control and Data Acquisition (SCADA).

In general, all grid-tied RE systems must be protected against grid-side issues, with abilities for active and reactive power control, voltage and frequency regulation, islanding, and restoration of grid connection during and after power outages. All of these can be done through advanced control strategies. Better forecasting techniques for the weather and load demands are also crucial to ensuring better control of the grid.

Issues Related to Grid Integration of Renewables

Integrating renewables or distributed generation units into the utility grids can disturb the system stability as it reduces the system inertia and affects the power quality. RESs are inherently intermittent and do not provide a constant power output, making forecasting and balance of demand and supply critically important. Integrating RESs into the utility grid can pose several threats to the grid, such as:

1. Voltage and Frequency Fluctuation: Due to the fluctuating nature of RESs, it is not possible to get a constant output from the RE generating unit, such as a solar panel or a wind turbine. The output varies to a great extent, which might cause the voltage and frequency levels to drop and rise suddenly, which is unwelcome in the utility grid and threatens grid reliability.
2. Power Fluctuations: The voltage fluctuation from the renewable energy generation units can also affect the output power. When integrated into the grid, these fluctuations might disrupt the entire grid by creating harmonics and affecting the grid power.
3. Power Quality: Power quality is defined as the property of the electricity grid to maintain specified conditions of voltage, power, frequency, and waveform. The presence of harmonics and the fluctuations in voltage and frequency are responsible for affecting the power quality of an electric system. Since the integration of RE into the utility grid causes voltage drops, frequency fluctuations, and harmonic content in the waveform, power quality might be compromised upon the injection of RE into the grid. Transients, long- and short-duration voltage variations and imbalance, waveform distortion, and frequency variations cause power quality issues resulting from renewable power injection into electrical grids.
4. Harmonics: RE integration into the utility grid has been found to increase the presence of harmonics in grid electricity caused by power electronic devices, or

inverters, used in grid integration of RE. The harmonic content can be mitigated by the use of devices such as filters and STATCOM.

5. Storage: The presence of storage devices in grid-tied RE systems enhances the flexibility and functionality of the system and greatly benefits the grid by mitigating the problem of RE intermittency. Appropriate control of the charging and discharging cycles of the storage system is necessary to ensure the proper running of the RE system and the reliable and stable delivery of clean electricity.

6. Optimal Placement of RESs: The RE units integrated into the utility grid must be optimally located and sized in the distribution network to minimize the loss of power and energy. This is a matter of concern in distributed generation. There are numerous algorithms developed for optimally placing the generating units.

7. Islanding: In case of any grid-side imbalances or disturbances, the RE system should be able to isolate itself from the utility grid in order to ensure self-protection. This is known as islanding and is accomplished by the inverters. When the problem has been solved, the RE system is reconnected to the grid either automatically or manually.

Many measures have been identified to mitigate or eliminate the issues related to grid integration of REs. Nevertheless, as long as the problem of intermittency of REs persists, the issues cannot be eliminated entirely. Spinning reserves are used in many utility grids to eliminate the issues that ruin the power quality. Moreover, the power quality in microgrids can be controlled with the use of custom power devices (CPD), such as an uninterruptible power supply (UPS), dynamic voltage restorer (DVR), transient voltage surge suppressor (TVSS), and distribution static synchronous compensator (D-STATCOM). In RE systems, the use of STATCOM is extensive for reactive power control purposes, improving transient conditions, removing voltage fluctuations, controlling voltage flickers, and damping harmonic components.

Case Study: A Grid-Tied Wind Farm The Biglow Canyon Wind Farm is an onshore wind power plant located in Oregon, USA. It was built in three phases and completed in 2010. It spreads over a vast area of 25,000 acres. It has 217 operational turbines, which generate net energy of 1069 GWh annually, working at a capacity of 27.1%. The cost of the wind farm was $1 billion. It can generate 450 MW of peak power, sufficient for 120,000 homes. The wind farm is grid-tied and supplies power to the high voltage (230 kV) transmission lines through a substation and a feeder transmission line (Table 10.1).

Table 10.1 Breakdown of the three phases of the Biglow Canyon Wind Farm and the turbine specifications

Characteristic	Phase 1	Phase 2	Phase 3
Commissioning	December 2007	September 2009	September 2010
Number of turbines	76	65	76
Turbine origin	Vestas	Siemens	Siemens
Turbine rating, MW	1.65	2.3	2.3
Diameter of turbine blades, m	82	93	93
Hub height, m	80	80	80
Total nominal power, MW	125.4	149.5	174.8

Each of the 217 turbines generates electricity at 600 V AC. A step-up transformer at the base of each turbine elevates the voltage to 34.5 kV. Since the turbines generate AC power, there is no need for an inverter to convert the DC voltage. From the transformer output, the power is fed to a central substation through underground cables buried at least 3 ft. below the ground. Fiber optic cables run along with the electric cables to transmit signals for the SCADA system that monitors and supervises the farm. The SCADA system performs both control and communication functions for each wind turbine and sends information to a central control computing system at the Operation and Maintenance (O&M) facility.

Exercise 10

1. Why is it necessary to integrate renewable electricity into the electricity grid?
2. Why is the study of grid integration of renewable energy necessary?
3. Discuss how converters play a role in the grid integration of renewable energy.
4. Why are energy storage systems required in grid-tied renewable energy systems?
5. Describe the role of net metering to boost the usage of renewable energy sources.
6. Explain why the grid integration of renewable energy is problematic.
7. Find out at least five European Policies regarding the integration of renewables into the utility grid.

Economic Aspects of Renewable Energy

11

Introduction

Renewable energy implementation is growing at an unprecedented rate as the world is transitioning toward adopting sustainable and greener technologies. However, a debate still continues regarding the cost of renewable energy generation compared to conventional forms of energy, mainly because of the infrastructure investment required for harvesting renewable energy, such as solar photovoltaics (PV), wind energy, and hydroelectricity. Although the initial installation of renewable energy harvesting mechanisms is expensive, once the payback period is over, the RESs are entirely free except for some regular operational and maintenance costs and can yield many more benefits than conventional energy sources.

From 2010 to 2019, the cost of all types of renewable energy has been dramatically reduced. The International Renewable Energy Agency (IRENA) reported that the cost of solar PV went down by 82%, concentrated solar power (CSP) by 47%, onshore wind by 39%, and offshore wind by 29%. In fact, many RE plants are now cheaper than the cheapest coal-based powerplant.

Table 11.1 shows the globally weighted average of the expected changes in investment costs and the levelized cost of energy (LCOE), along with the increase in the capacity factor of various RE technologies through the years 2015 to 2025. This table indicates that the cost of RE technologies is gradually coming down, while the capacity of these powerplants is on the rise. The cost of RE technologies cannot be determined linearly because it depends on many factors. Some basic terminologies are commonly associated with RE economies. To get an idea about the RE economies, they must be understood first. These terminologies are defined subsequently.

Net present value (NPV): The NPV is the present value of the future income after deducting the value of the investment costs and other future expenditures. A positive NPV indicates that the financial justification and viability of the project.

© The Author(s), under exclusive license to Springer Nature
Switzerland AG 2021
E. Hossain, S. Petrovic, *Renewable Energy Crash Course*,
https://doi.org/10.1007/978-3-030-70049-2_11

Table 11.1 Global Weighted Average of the rate of change in the capacity factor, investment costs, and LCOE of solar and wind power technologies over the years from 2015 to 2025 [1]

Technology	Capacity factor			Investment costs ($/kW)			LCOE ($/kWh)		
	2015 (%)	2025 (%)	Percent increase (%)	2015	2025	Percent reduction (%)	2015	2025	Percent reduction (%)
Solar PV	18	19	8	1810	790	57	0.13	0.06	59
Parabolic Trough Collectors	41	45	8.4	5550	3700	33	0.15	0.09	37
Solar Power Tower	46	49	7.6	5700	3600	37	0.15	0.08	43
Onshore Wind	27	30	11	1560	1370	12	0.07	0.05	26
Offshore Wind	43	45	4	4650	3950	15	0.18	0.12	35

Internal rate of return (IRR): It is a discount rate that allows the NPV to become zero. IRR helps to determine whether a project is financially profitable or not. Therefore, a higher value of IRR is preferable in the case of any project.

Compound annual growth rate (CAGR): It is a measure of the growth of sales, revenue, or income. CAGR provides a long-time perspective about the historical financial performance of a company/business or its expected future performance. It is measured as a percentage and merely tells us the yearly constant growth rate or rate of return after a definite period based on the initial and final values. It cannot reveal whether the growth rate was indeed uniform throughout the period or not. CAGR can be calculated from the equation given subsequently:

$$\text{CAGR} = \left(\frac{V_f}{V_i}\right)^{\frac{1}{t}} - 1, \tag{11.1}$$

where, V_i and V_f are the initial and final values of the project, and t is the time in years.

Payback Period: It is the time by which the initial investment costs are earned back.

Levelized cost of energy (LCOE): The LCOE is the ratio of the NPV of the total installation and operation costs of the technology to the total electricity generation over the system's entire lifetime. The equation to calculate the LCOE of RE technologies is:

$$\text{LCOE} = \frac{\sum_{t=1}^{n} \frac{I_t + M_t + F_t}{(1+r)^t}}{\sum_{t=1}^{n} \frac{E_t}{(1+r)^t}}, \tag{11.2}$$

The symbols in the equation are explained subsequently:
I_t = cost of investment in the year t.
n = economic lifetime of the system.
r = rate of discount, or internal rate of return (IRR).
E_t = generated electricity in the year t.
F_t = fuel cost in the year t.
M_t = operation and maintenance (O&M) costs in the year t.

PV System Costs

The worldwide solar capacity has grown from only 40 GW in 2010 to 580 GW in 2019, with a growth factor of 14 [IRENA]. This increase has been coupled with a decrease in module prices by 90%. For large-scale solar, the levelized cost of energy (LCOE) in 2020 was $0.068 per kWh, in contrast to $0.378 in 2010. The LCOE fell by 13.1% between 2018 and 2019.

Table 11.2 Falling average costs of large-scale solar in major regions of the world during 2010–2019 [2]

Country	Percentage of price decline (%)
United States	66
Germany	73
France	77
Australia	78
Spain	81
China, Italy, South Korea	82
India	85

The price of solar PV modules has always been on the decline. Swanson's Law recognizes that the solar PV module price reduces by 20% for each twofold increase in the aggregated shipped volume of the modules. At the present rate of utilization of solar modules, the costs decreased by 75% in every 10 years. The cost of crystalline Si PV cells per Watt reduced from US$76.67 in 1977 to only US$0.21 in 2020—a sharp reduction by more than 99% in 43 years only. The decline of solar costs in several countries is listed in Table 11.2.

The reasons for this cost decrease are rapidly declining manufacturing costs, investments for large-scale projects, and government incentives. For rooftop, grid-tied systems, the cost breakdown shows the following: module—60%, inverter—10%, the balance of plant (all other electrical components and installation)—23%, and engineering and procurement—7%.

Figure 11.1 represents the growth of the solar PV deployments in terms of installed capacity in the United States, along with the exponential decline in the average prices per unit power. Similarly, the costs of all other RE technologies can be lowered by large deployments.

Wind Power Costs

In this section, we shall analyze the costs of generating wind power at a large scale or utility scale. The cost analysis for a wind farm takes into consideration several factors, such as the suitability of the site for the wind farm, which depends on the wind speeds and flow variations at that site, the type and size of the turbines, the tower height, the spacing between adjacent turbines, nearby localities and their tolerance with the noise and visual obstruction, etc. All these technical and nontechnical factors eventually reveal whether the project is cost-effective or not. The cost analysis, although it will vary with project site and size, includes several types of expenditure, namely:

1. Investment or capital costs.
2. Operation and maintenance (O&M) costs.
3. Land lease or purchase costs.
4. Costs for compensating the impact on life, environment, and other technologies.

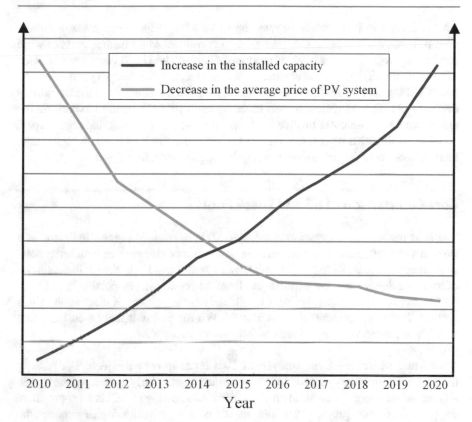

Fig. 11.1 The price of solar PV systems is gradually falling over the years, while the annual installations are increasing gradually [3]

5. Project financing structure (equity and debt finance).
6. Clean development mechanism (CDM) up-front and administrative costs.

Expenditures related to wind power projects include but are not limited to the assessment of the wind resource, site selection, type of plant (onshore or offshore), feasibility studies, cost of construction materials and labor, government permits, legal and consultation fees, interconnection studies, repair, replacements and upgrades, operation and maintenance, warranty and insurance, protection and metering arrangements, taxes, etc.

The total costs for the installation of wind turbines depend on the number of turbines and project location. Small-scale projects cost more per installed kW of generation compared to large-scale projects. Installation of small-scale or residential wind turbines might cost from $3000 to 8000/kW below 10 kW ratings, whereas for turbines above 10 kW, the costs might increase to $50,000–80,000/kW. The cost can range from $1.3 to 2.2 million per MW in the case of utility-scale wind turbines. The typical initial cost of large-scale wind turbines is $1.3 million per MW of electric power generation capacity. Most commercial turbines are rated at 2–3 MW, while offshore turbines might be rated up to 12 MW. With the increase in turbine size, number, and tower height, the cost also increases. Maintenance costs of wind turbines are usually 1–2 ¢/kWh of produced energy. The O&M costs are estimated to be around $48,000/MW of electric power produced in the first 10 years of operation. O&M costs usually increase with the age of the installations as the machinery becomes old. The depreciation costs saved up each year can be used to perform the O&M works. These costs involve land rent, insurance, servicing and repair, spare parts, administrative works, etc. The cost of offshore turbines is much higher than their onshore counterparts since they require added protection systems.

Cost Comparison of RE and Fossil Fuels

The cost per unit of energy produced, i.e., $/kWh, is an important indicator of a system's cost-effectiveness and a reliable parameter of comparison between renewable energy sources (RESs) and conventional fossil fuels. From the ecological point of view, RESs are the clear winners, but from an economic perspective, whether or not RESs are cost-competitive with fossil fuels is the subject of discussion in this section. Table 11.3 presents the LCOE in $/kWh for different fuel types, including fossil fuels, nuclear power, and renewable energy sources.

Case Study: Lifetime Cost Analysis of Two Hydropower Projects For judging the cost–benefits of any renewable energy project, particularly in comparison to fossil fuel-based energy generation projects, life cycle costing (LCC) is an appropriate way. Hydropower projects are the most expensive technologies among the renewables due to their massive scale of construction and installations. Yet, hydropower accounts for the maximum share of electricity generation from clean sources

Table 11.3 Cost comparison of various electricity generation sources in $/kWh [US EIA statistics and analysis from Annual Energy Outlook 2019]

Fuel type	Type of power plant	Cost or LCOE, cents/kWh
Fossil fuels	Coal with carbon capture and sequestration (CCS)	12–13
	Combined cycle natural gas	4.3
	Combined cycle with Carbon Capture and Sequestration (CCS)	7.5
Nuclear	Nuclear	9.3
Renewables	Onshore wind	3.7
	Offshore wind	10.6
	Solar thermal	16.5
	Solar photovoltaic	3.8
	Geothermal	3.7
	Hydro	3.9
	Biomass	9.2
	Fuel cells	12.7

Table 11.4 Breakdown of the percentage share of different cost heads in the total costs associated with a grid-tied hydropower plant [4]

Type of cost	Project component	Percentage of total costs		
		Minimum (%)	Average (%)	Maximum (%)
Installation or capital costs	Civil (construction and buildings) work	17	45	65
	Mechanical equipment buying and installing	18	33	66
	Integration to the grid	1	6	17
	Planning and design	6	16	29
	Land cost or rent with taxes	1	3	18
Operation and maintenance (O&M) costs	Operation costs	20	51	61
	Salaries of engineers, technicians, and laborers	13	39	74
	Miscellaneous costs	5	16	28
	Materials for repair, replacement, or additional construction	3	4	4

and is one of the cheapest electricity sources. The total investment costs of hydro-power plants vary depending on the location, availability of natural reservoir, and existing dam or construction of an artificial reservoir and a new dam, design, cost of materials, and labor cost. Typically, the cost of large-scale hydro projects ranges from $1000 to 3500/kW. IRENA presents a breakdown of the total costs into the installation, or capital costs (also known as Capital Expenditure or CAPEX), and operation and maintenance (O&M) costs in Table 11.4. Table 11.5 presents a comparative picture between the Grand Coulee Dam and the Hoover Dam, two of the most expensive but most successful projects in US history.

Exercise 11

1. Briefly opine whether renewable energy technologies are economically viable or not.
2. Prepare a rough cost analysis of a 5-kW residential solar PV rooftop project with a suitable storage system.
3. Find out the trend of the costs of energy storage systems, particularly Lithium-ion batteries, over the years from 2000 to 2020.
4. What caused a decrease in the price of silicon solar cells?
5. How can the costs of renewable energy technologies be reduced?

Table 11.5 Comparative picture of the Grand Coulee Dam and the Hoover Dam [5]

Points to note	Grand Coulee Dam [6, 7]	Hoover Dam [8, 9]
Construction started in	July 16, 1933	1931
Opening date	1942	1936
Location	County: Grant/Okanogan State: Washington	County: Clark, Mohave State: Nevada, Arizona
Operator	US Bureau of Reclamation	US Bureau of Reclamation
Surface area, km^2	324	640
Catchment area, km^2	191,918	435,000
Installed capacity, MW	6809	2080
Capacity factor, %	36	23
Annual generation, TWh	20.24	4.2
Total number of turbines	33	17
Construction costs at the time of construction	$163 million in 1943 and $730 million in 1973	$49 million in 1931
Construction costs in 2019-dollar value	$5.23 billion	$675 million
Rate of electricity produced, ¢/kWh	Residential: 7.44 Commercial: 6.77 Industrial: 6.01	Residential: 8.19 Commercial: 10.66 Industrial: 6.48
Construction costs repaid by	Not yet	1987
Expected lifetime of the dam	10,000 years	10,000 years

References

1. IRENA. The power to change: solar and wind cost reduction potential to 2025, June 2016, ISBN 978-92-95111-97-4 (2016)
2. https://www.pv-magazine.com/2020/06/03/solar-costs-have-fallen-82-since-2010/
3. https://www.seia.org/solar-industry-research-data
4. IRENA. Renewable Power Generation Costs in 2019, June 2020. ISBN 978-92-9260-244-4 (2020)
5. https://www.usbr.gov/pn/grandcoulee/pubs/factsheet.pdf
6. https://www.usbr.gov/pn/grandcoulee/about/faq.html
7. https://www.electricitylocal.com/states/washington/grand-coulee/
8. https://www.usbr.gov/lc/hooverdam/faqs/powerfaq.html
9. https://www.electricitylocal.com/states/nevada/boulder-city/

Challenges of Renewable Sources of Energy

12

Introduction

Every advancement of science and technology has had an adverse effect on the environment. Fossil fuels are easily available and cheap sources of energy, but they are responsible for global warming and climate change.

The current dependence on fossil fuels is temporary and alternative, and sustainable energy sources must be introduced for primary energy needs. However, there are certain problems associated with RE technologies that impede their progress and make investors reluctant to invest in them. First, RESs are intermittent and unpredictable. For example, the amount of solar power is maximum during the afternoon, the wind flow is the strongest at night and in coastal areas, geothermal energy is not uniformly available everywhere, and hydroelectricity cannot be produced without a specific location and supporting infrastructure. Second, the high capital costs and the long payback periods associated with RESs lead to hesitation by the investors in funding RE projects.

Apart from the intermittency and uncertainty of RESs, some prominent negative sides of renewable energy technologies include the impact on the environment, high costs, and the problem of effective waste management. The following paragraphs detail these challenges that RESs have yet to overcome in order to replace conventional energy sources.

Intermittency of Renewables

The most dominating problem with RESs is their intermittent and unpredictable nature. The amount of power that can be extracted from RESs varies with the time of the day, season, location, and several other factors.

Hydroelectricity is the largest contributor to energy storage systems worldwide, but it is also dependent on the seasons. Peak production can be obtained during the

© The Author(s), under exclusive license to Springer Nature
Switzerland AG 2021
E. Hossain, S. Petrovic, *Renewable Energy Crash Course*,
https://doi.org/10.1007/978-3-030-70049-2_12

rainy season when the water levels are high, and the production reduces during winter.

Solar power, both photovoltaic and thermal, solely depends on the insolation at the site and is disrupted due to clouds and shading. It is entirely absent from sunset to sunrise and reaches the peak only during midday.

Wind power depends on wind speed and direction, which varies depending on location. The amount of underground geothermal energy is also nonuniform across various regions of the Earth. Ocean energy, i.e., wave and tidal energy, depends on the occurrence of water waves and tides, which are natural phenomena utterly dependent on the motion of the Earth and the moon and atmospheric conditions.

Bioenergy is available as biomass and biofuels. It is comparatively reliable and available throughout the year and in all places, while the large variety of raw materials ensures continual supply. However, bioenergy is not wholly a clean energy source since it is responsible for carbon dioxide emissions.

Impact of Renewables on Living Beings

Renewable energy sources are not always benign toward living beings such as birds and animals. Humans are also affected but to a less severe extent. RESs are designed with the prime focus on creating a green, carbon-free system, but the concern for animals and birds is sometimes neglected. Solar power towers concentrate sunlight to produce immense heat, which can burn birds and insects.

Wind turbines are responsible for the death of birds and bats caused by a direct collision. However, recent studies have shown that painting one blade of the turbine black can significantly drop the bird and bat mortality rates.

Hydropower dams are responsible for changing the flow of rivers or lakes, which obstructs the free movement of fish and disrupts the natural life cycle of aquatic animals, such as the endangered salmon in the Pacific Northwest and stingray in Southeast Asia.

Many forest areas have to be cleared to collect raw materials for producing bioenergy; this meddles with the natural habitat of many forest species, and they are forced to relocate.

Major solar projects are designed in desert areas, which are home to many rare species. Solar projects have forced the Giant Kangaroo Rat of the Carrizo Plain and the Desert Tortoise of the Mojave Desert to be listed as endangered.

These are all potential negative impacts of RE projects on living beings, which can be ameliorated by thorough preplanning and proper rehabilitation of the threatened species.

Impact of Renewables on the Environment

In contrast to energy generation from fossil fuels, renewable energy is clean, nonpolluting, and emission free. However, RESs are sometimes responsible for causing harm to the natural environment, although the type and intensity of the impact vary based on the technology, size of the project, the location, surroundings, and a number of other factors.

RE projects take up a vast amount of land, reducing the available land for agricultural and residential uses. These projects also cause loss or relocation of wildlife and destruction of the ecosystem at the selected site. Wind and solar technologies often use toxic chemicals like plastic, heavy metals (Pb, Cd, Sb), etc., which can harm the environment when in contact with the ground or emitted into the air.

Biomass and biofuels are already controversial because they emit CO_2 and other toxic gases similar to fossil fuels. The only difference is that bioenergy is renewable while fossil fuels are depleting. Many trees are also cut down to supply raw materials for producing bioenergy.

Large quantities of freshwater are also utilized in renewable technologies, which poses a threat to the availability of drinking water, although the measures are taken to use wastewater instead of freshwater.

The excavation and construction work for geothermal and hydroelectric projects cause massive geological and ecological changes. Offshore wind and ocean power projects may significantly disturb the aquatic ecosystem.

On a positive note, most of the environmental impacts of RESs can be mitigated by proper planning, siting, and sizing of RE plants. Prevention is better than cure; therefore, systematic planning and analytical feasibility studies can help mitigate the environmental impacts of renewables.

Renewable Energy Systems Waste Disposal

Renewable energy is clean compared to fossil fuels, but some technologies leave behind toxic wastes that are difficult to dispose. Solar panels contain toxic elements, such as Cd, and have been found to emit toxic pollutants. The main constituent in most solar panels is glass, which contains plastics, Pb, Cd, and Sb, making it unsuitable for recycling. When disposed of in landfills, the solar panels become a severe problem at the end of their useful life, leaching toxic chemicals and also causing a waste of valuable resources such as Si, Al, and Cu.

In 2018 alone, solar panels have produced 250,000 metric tonnes of waste, and the amount could be 78 million metric tonnes by 2050. However, better recycling techniques and strong regulations to recycle solar panels can transform solar photovoltaics into truly clean technology.

Wind power is very cheap and free of emissions. Although 85–90% of the turbine components can be recycled or reused, such as steel, copper wires, electronics, and gearing, the decommissioned fiberglass blades are difficult to dispose of and might create 43 million tonnes of waste by 2050. They are huge in size, so they are crushed before getting dumped into landfills—which is technically challenging and expensive. Therefore, extensive research is needed to develop methods for recycling, reusing, or repurposing the waste created by renewable energy technologies.

Cost of Renewables

The high capital cost of RESs is the dominating factor that has held back the complete flourish of renewables, although the fuel cost is zero and the operation and maintenance costs are minimal. In 2017, the average cost of installing solar systems was about $2000/kW for large-scale systems and around $3700/kW for residential systems.

Wind turbines cost around $1200–1700/kW. In contrast, natural gas plants cost only around $1000/kW, which is more attractive. Although the cost of RESs has indeed been decreasing over the years, as seen from the previous chapter, they are still not cost-competitive compared to fossil fuels. A study by Wood Mackenzie revealed that it would cost $4.5 trillion for the United States to switch to a 100% renewable-powered future by 2030. This cost not only involves the capital costs of building new solar and wind power facilities but also the costs of new transmission lines, power translators, and energy storage systems. Although renewables are expensive to install, their lifetime cost is very cheap compared to that of fossil fuels. In 2017, the lifetime cost of large-scale solar was $43–53/MWh and wind power was $30–60/MWh. On the other hand, the lifetime cost of a natural gas-powered plant was $42–78/MWh and of a coal-fired plant was $60/MWh (Table 12.1).

Case Study: Failure Versus Success of Solar Thermal Plants in the United States This case study juxtaposes the stories of two solar thermal plants in the United States, one being delayed despite once being the largest and most expensive solar project in the United States, and the other being an example of a successful concentrated solar power (CSP) plant that has been generating millions of dollars in revenue ever since its commission. Table 12.2 presents a comparative picture of the two solar power plants.

The Crescent Dunes Solar Thermal Plant was a huge investment funded by the US Department of Energy and Citigroup, yet it failed to deliver its expected output energy. The plant suffered a leak in its molten salt storage only months after its commissioning, which led to an 8-month shutdown. Meanwhile, solar PV costs plummeted to a rate far lower than solar thermal, eventually making investors look away from solar thermal and get interested in solar PV.

Another factor contributing to the downfall of the Crescent Dunes plant was its low-capacity factor, which is the ratio of the amount of energy generated over a period of time to the maximum generation capacity. While other competing technologies increased their capacity factors, Crescent Dunes could not achieve even half of its planned capacity factor. As a result, the project's cost per unit power generated rose, making it financially unattractive.

The Mojave Solar project used parabolic trough technology, which is more common and technologically advanced compared to the solar power tower technology used in the Crescent Dunes plant. Despite having higher capital investments, the Mojave project generated enough electricity to reduce the capital cost per unit power generated to lower than that of the Crescent Dunes plant. Both plants began

Table 12.1 Challenges of different forms of renewable energy and their probable solutions.

Renewable energy technology	Challenges	Probable solutions
Hydroelectricity	1. Large-area requirement 2. Disruptions to aquatic life 3. Altering the natural flow of water bodies	1. Building hydroelectric projects in hilly areas to make use of the natural elevation 2. Proactive planning to ensure that the natural species are least affected, and if possible, arrange for rehabilitative alternatives 3. Proper and thoughtful siting 4. Micro-hydro facilities have a lesser impact than large hydroelectric plants
Wind power	1. Death of birds and bats 2. Noise created by the turbines 3. Appropriate locations are far from power consumption sites 4. Turbine blades create waste	1. Painting one turbine blade 2. Selecting the proper location and height of the turbines 3. Manufacturing turbine blades from recyclable, reusable, or repurposable materials
Ocean power	1. Complexity of harnessing wave power; Design inefficiency 2. No practical implementations and reluctance in investments 3. Threat to aquatic life	1. Researching and developing efficient methods of harnessing the energy from tides and waves 2. Running extensive simulations 3. Proper planning to ensure aquatic life is not harmed
Biomass	1. Deforestation and loss of animal habitat 2. Emissions 3. Takes a lot of time to regenerate the biomass	1. Using naturally fallen or uprooted trees instead of cutting down fresh trees 2. Using crop leftovers 3. Planting a lot of trees to compensate for the trees lost 4. Creating alternative and cheap carbon sequestration methods
Biofuels	1. Energy crops compete for the land with food crops 2. Emissions 3. Production of biofuels requires special techniques and processes	1. Proper land choice 2. Prioritizing food crops over energy crops 3. Making the emissions carbon neutral, if not carbon negative
Geothermal	1. Complex infrastructure 2. Altering the structure of the Earth 3. Vulnerability to earthquakes and natural disasters	1. Utilizing the naturally available geothermal hot springs and geysers, instead of intensive digging and excavating 2. Using old and unused caverns 3. Focusing more on geothermal aquaculture to make geothermal heat extraction more economical and efficient

(continued)

Table 12.1 (continued)

Renewable energy technology	Challenges	Probable solutions
Solar thermal	1. Creating heating centers by the positioning of the mirrors which can burn insects and birds 2. Large upfront costs	1. Proper positioning and siting 2. Taking appropriate measures to keep birds away from the heating centers created
Solar photovoltaics	1. High upfront costs 2. Cost of semiconductors 3. Usage and mining of important chemicals such as Al, Cu, Cd, In 4. Toxic Wastes 5. Plastic pollution	1. Using more solar panels in order to bring down the costs 2. Recycling, reusing, or repurposing the panels or parts of the panels 3. Extracting the useful elements from the panels to reuse them for other purposes
Energy storage	1. Toxic chemicals from batteries 2. Mining elements from the Earth 3. Hydrogen storage—hydrogen production is quite expensive and might be a cause of hazards if the stored hydrogen leaks out	1. Prioritizing rechargeable batteries 2. Second life of batteries 3. Focusing on cleaner methods of energy storage, such as mechanical, thermal, and electrical energy storage

Table 12.2 Comparison between the Crescent Dunes Solar Energy project and the Mojave Solar project

Topic of comparison	Crescent dunes solar energy project	Mojave solar project
Start of construction	2011	2011
Commission date	2016	2014
Location	Nevada	Mojave Desert, California
Developer	SolarReserve	Abengoa
Capacity, MW	110	250
Annual generation	196 GWh in 2018	582 GWh
Solar farm type	Concentrated solar power	Concentrated solar power
Technology	Solar power tower	Parabolic trough
Site resource, kWh/m²/year	2685	2685
Site area, acre	1670	1765
Total collector area, acre	296	385.323
Capacity factor	Planned: 51.9%; Actual: 20.3%	27.7%
Current status	Obsolete	Operational
Energy storage	1100 MWh$_e$	Absent
Construction cost	$0.975 billion	$1.6 billion
Capital cost per unit	$8.86 million per MW	$6.4 million per MW

construction in the same year, but they did not maintain the same progress and technological advancements. Crescent Dunes also took more time for construction, so the two extra years gave Mojave a kickstart in revenue collection. The salt leak accident and the consequent shutdown were also big blows to the Crescent Dunes plant. Even if Crescent Dunes could be restarted, no one would be ready to invest or to take control of this compromised project.

Exercise 12

1. Explain three main challenges or problems of solar photovoltaic panels.
2. How do you think the problem of the relocation of animals due to the establishment of renewable energy farms (such as wind farms) can be solved?
3. Make a list of the species endangered and threatened by the expansion of renewable energy technologies and reflect on how they can be preserved. Also, shed some light on how renewable energy technologies can flourish, making sure that animals can be rightfully preserved.
4. Discuss some ways to ensure the safe disposal of waste from renewable energy technologies.

Comparative Study of Renewable Sources of Energy

13

Introduction

Renewable energy sources are presently used all over the world. Some areas have a natural inclination to a particular source because of geography, economy, and available infrastructure. A comparative analysis of the different types of renewable energy can help fully explore the potential of renewable energies worldwide.

Table 13.1 compares seven basic types of renewable energy technologies: hydropower, wind power, ocean power (tidal and wave power), bioenergy (biomass, biofuel, and biogas), geothermal power, solar photovoltaics, and solar thermal. Each type of RE technology is suitable for a specific purpose, location, load size, etc. Not all types can be used for the same purpose, nor can all purposes be satisfied by a particular type. Hence, while choosing a type of technology for a project, one must conduct thorough research and feasibility analysis to justify which RE technology will work best for the specific purpose and location.

Case Study: Stillwater Triple Hybrid Renewable Energy Powerplant The Stillwater Triple Hybrid Powerplant in Fallon, Nevada, USA, is a unique renewable energy powerplant as it incorporates three different RE technologies that work together to produce 61 MW of electric power based on the Organic Rankine Cycle. The plant has a 33 MW geothermal plant, a 26 MW solar PV plant, and a 2 MW solar thermal or concentrated solar power (CSP) plant.

This is the world's first combined RE powerplant and was constructed in 2001. Initially, it was only a geothermal binary powerplant based on the Organic Rankine Cycle, which came online in 2009. It was a four-turbine unit with isobutane as the working fluid that expanded and rotated the turbines. Then, 89,000 solar PV panels were added in 2012 to maintain electricity production when geothermal resources could not be sufficiently utilized during the hot days. In the hybrid system, the solar PV and geothermal plants complimented each other perfectly and produced a nearly

E. Hossain, S. Petrovic, *Renewable Energy Crash Course*, https://doi.org/10.1007/978-3-030-70049-2_13

Table 13.1 A comparative analysis of different renewable energy technologies

Technology	Hydro	Wind	Ocean	Bioenergy	Geothermal	Solar PV	Solar thermal
Rank among renewables in terms of energy produced	1st	2nd	7th	4th	5th	3rd	6th
Contribution to global energy produced, %	7	2	<1	10	0.2	>1	>1
Contribution to global electricity produced, %	16	5	<1	2	0.3	>2	>1
Contribution to global renewable energy, %	~30	10%	0.005	~50	~4	<5	<5
Contribution to global renewable electricity, %	~60	18	0.02	9	1	7	<5
Location and geography dependence	Yes	Yes	Yes	No	Yes	Yes	Yes
Seasonal dependence	Yes	Yes	Yes	No	No	Yes	Yes
Energy production (2018 or 2019 data), TWh	4333	Offshore: 67; Onshore: 1323	1.2	589	92.7	720	15.6

Technology	Hydro	Wind	Ocean	Bioenergy	Geothermal	Solar PV	Solar thermal
Installed generation capacity, GW	1308	563	0.535	117	13.2	512	6.45
Origin of the energy source	Solar radiation	Solar radiation	Lunar and solar gravity; solar radiation	Plant decay	Heat from the formation of Earth and radioactive isotope decay in the core	Solar radiation	Solar radiation
Primary origin	Hydrological cycle	Wind flow, due to differences in temperature caused by sunlight	Waves, tides, and thermal gradient due to sunlight	Photosynthesis	Convection and conduction	Sunlight	Sunlight
Stored energy type	Gravitational potential energy	Kinetic energy	Kinetic and thermal energy	Chemical and thermal energy	Heat energy	Light energy	Thermal energy
Governing equation	Potential energy, $E_P = mgh$; m = mass of water; g = acceleration due to gravity, h = height of reservoir	Betz Formula; $$P_{max} = \frac{16}{27}\frac{\rho}{2}V_1^3\frac{\pi D^2}{4}$$ Watt	Tidal Barrage: $PE = \rho r A R^2/2\ T$; Wave Energy: $P = 2A^2\ T$	No specific equation since different biomaterials produce energy undergoing different types of chemical reactions	$Q/t = k(T_h\text{-}T_e)/d$; k = thermal conductivity	Energy at a specific frequency, $E = hU$; E = energy; h = Planck's constant; U = frequency (Hz)	Radiative Thermal Equilibrium Equation: $(1 - a)S = 4\varepsilon\sigma T^4$; Emissivity ($\varepsilon$) and Stefan-Boltzmann Constant (σ)

(continued)

Table 13.1 (continued)

Technology	Hydro	Wind	Ocean	Bioenergy	Geothermal	Solar PV	Solar thermal
Year of initiation of the technology	Sixth century B.C.	Ancient Egypt, 5000 years ago	Tidal: 1100 A.D.	Since the beginning of human history	Nearly 10,000 years ago	1883, by Charles Fritts	1000 A.D. Mesopotamia
Energy content (if applicable)	$P = 10\eta QH_e$; η = efficiency, Q = flow rate, H_e = effective head	$P = \frac{1}{2}\pi r\rho Av^3$	$P = 2A^2T$	~5.3% of solar energy	Maximum heat flow: 300 mW/m³	$P = I \times V$; P = power; I = current; V = voltage	Solar constant: 1367 W/m²
Energy conversion type/mode	Gravitational potential to kinetic to electrical	Kinetic to electrical	Kinetic to electrical (tidal); solar heat to kinetic to electrical (OTEC)	Solar light to chemical to heat to electrical	Heat to kinetic to electrical	Solar light to electricity	Solar heat to kinetic to electrical
Device used for conversion	Fourneyron's Turbine, Francis Turbine, Impulse Turbine incorporated in Dams	Wind Turbines	Wells Turbine, OWC, TAPCHAN, Pendulum Device, Floating or Buoy Devices, OTEC	Steam Turbine, Gas Steam Turbine, Combined Cycle, Gas Engine, Distillery	Steam Turbines, Dry Steam Turbines, Flash Steam Turbines, Binary Cycle, Hot Rock Plants	Photovoltaic Solar cells	Solar Thermal Collector, Parabolic Troughs, Thermosyphon Solar Water Heaters, Flat Plate Collectors, Power Tower, Solar Chimney
Conversion efficiency, %	90–95	59.3 (maximum)	80	20–25 (electrical); 75–80 (CHP)	12	25	15–20

Technology	Hydro	Wind	Ocean	Bioenergy	Geothermal	Solar PV	Solar thermal
Conversion limitation	Topographical or geographic	Weibull distribution	Tidal force, coastal land distribution	Energy content of biomass materials	Radiative heat from Earth's core	Thermal Properties of semiconducting materials and total irradiance	Earth's radiative equilibrium
Range of size of the device	Hydro plants are incredibly variable. They can be as small as a few kilowatts, to as large as 22,500 MW	Wind turbines are very scalable. Residential turbines can be as small as a few kW. The largest turbine is 12 MW	1.2–254 MW	4–1500 MW	5–1520 MW	Each solar cell can produce only 1–2 Watt. Solar PV is also very scalable	Parabolic Trough: 30–320 MW Power Tower: 10–200 MW Dish/engine: 0.005–0.025 MW
Largest device size	22.5 GW (Three Gorges Dam, China)	20 GW (Gansu Wind Farm, China)	254 MW (Sihwa Lake Tidal Power Station, South Korea)	740 MW (Ironbridge biomass power plant, Severn Gorge, UK)	1.2 GW (Geysers Geothermal Complex, California, USA)	2.245 GW (Bhadla Solar Park, India)	580 MW (Noor Ouarzazate Solar Complex, Morocco)
Capital investment ($/ kW)	1000–3500	2000	117–382 (Based on Sihwa and La Rance)	Around 4000	2500	1000–2000	4000–8000

(continued)

Table 13.1 (continued)

Technology	Hydro	Wind	Ocean	Bioenergy	Geothermal	Solar PV	Solar thermal
Operational and maintenance costs per year	2–2.5% of investment costs	20–25% of the LCOE of total kWh produced each year	$0.43–0.65/kW	Fixed: 2% of capital costs; Variable: $0.005/kWh	$0.01–0.03/kWh	$4–9.5/kW(DC)	$0.02–0.04/kWh
Weighted average cost of generating electricity, ($/kWh)	0.02–0.27	0.053 (onshore); 0.115 (offshore)	0.61	0.05–0.09	0.07	0.068	0.182
Energy return on investment (EROI)	84	18	15	1.2–1.6	6.6	6.8	2.1–12
Lifetime (generating device), years	50+	20–25	20–120	20–30	20–30	25–30	25–30
General hazard or catastrophic level	Dams can cause environmental issues. Sedimentation issues, ecological issues, water chemistry changes, large land use and inundation of people	Electrical hazards for workers. Collapse of the turbine or its blades	Potential environmental impact	Some occupational hazards such as CO emission	Some occupational hazards such as hydrogen sulfide emission	None	None

Technology	Hydro	Wind	Ocean	Bioenergy	Geothermal	Solar PV	Solar thermal
Emission of CO_2	None	None	None	0.046 kg/kWh	0.122 kg/kWh	None	None
Lifetime emission of CO_2	None	None	None	10,074 kg	26,718 kg	None	None
Emission of SO_x	None	None	None	0.03 kg/kWh	None	None	None
Emission of NO_x	None	None	None	0.00017–0.00094 kg/kWh	None	None	None
Other environmental concerns	Affects aquatic biodiversity	Turbines are a huge waste problem; bird and bat deaths	None	Deforestation; competing with food crops for fertile land	Pollution of groundwater aquifers	Toxic wastes	Bird deaths

linear output throughout the 24-h span—the daytime dominated by solar and the nighttime dominated by geothermal.

Later in 2015, a CSP plant using linear parabolic trough technology was added so that the geothermal reservoir could be further heated from the solar heat, maximizing the efficiency of the geothermal plant. The CSP plant increased the overall output by roughly 3.6%. The total investment for this triple hybrid powerplant was estimated to be only $15 million. It generates 33 million kWh of energy annually and produces about 200 GWh of electricity per year, with 150 GWh from geothermal, 44 GWh from solar PV, and 3 GWh from solar thermal or CSP. Thus, the geothermal and solar PV complemented each other to produce nearly linear output power. The addition of the solar thermal plant to the geothermal plus solar PV unit helped to elevate the output level by a significant amount.

Exercise 13

1. Suppose you are to design a 50 MW power generation plant at a location close to your house. Which renewable resource would you use and why? Write the name of the place you selected and analyze your choice based on the climatic and topographic features of the selected place.
2. Briefly compare solar PV and solar thermal technologies.
3. Briefly compare tidal and wave power plants.
4. What costs are included in the Operation and Maintenance costs of a RE project?
5. Explain how solar and wind technologies might complement each other in a solar-wind hybrid power plant.

Index

Printed in the United States
by Baker & Taylor Publisher Services